# MRE

Materials Research and Engineering
Edited by B. Ilschner and N.J. Grant

B.I. Medovar   G.A. Boyko
Editors

# Electroslag Technology

**With 129 Figures**

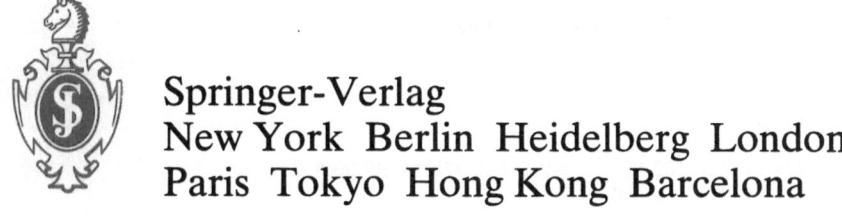

Springer-Verlag
New York  Berlin  Heidelberg  London
Paris  Tokyo  Hong Kong  Barcelona

**B.I. MEDOVAR**

E.O. Paton Electric Welding Institute
Kiev, Ukraine
USSR

**G.A. BOYKO**

E.O. Paton Electric Welding Institute
Kiev, Ukraine
USSR

*Series Editors*

**Prof. BERNHARD ILSCHNER**

Laboratoire de Métallurgie Mécanique
Département des Matériaux, Ecole Polytechnique Federale
CH-1007 Lausanne/Switzerland

**Prof. NICHOLAS J. GRANT**

Department of Materials Science and Engineering
Massachusetts Institute of Technology
Cambridge, MA 02139/USA

Library of Congress Cataloging-in-Publication Data
Ėlektroshlakovaiа tekhnologiiа. English.
    Electroslag technology/[edited by] B.I. Medovar, G.A. Boyko;
translated by Olga Gavrilovic.
        p.   cm.
    Translation of: Ėlektroshlakovaiа tekhnologiiа.
    Includes bibliographical references.
    1. Electroslag process.   I. Medovar, B.I. (Boris Izrailevich)
II. Boĭko, Georgiĭ Aleksandrovich.   III. Title.
TN685.5.E4E5413   1990
669'.028'4—dc20                                    90-36464

Printed on acid-free paper.

© 1991 Springer-Verlag New York, Inc.
Softcover reprint of the hardcover 1st edition 1991

Typeset by Asco Trade Typesetting Ltd., Hong Kong.


9 8 7 6 5 4 3 2 1

ISBN-13:978-1-4612-7762-0      e-ISBN-13:978-1-4612-3018-2
DOI: 10.1007/978-1-4612-3018-2

# Preface

Dr. Boris Medovar, a member of the Soviet Academy of Sciences, is a prominent member of the E.O. Paton Electric Welding Institute in Kiev, one of the pre-eminent institutes of the USSR. The Paton Institute, internationally famous for its entrepreneurial efforts in electrical welding processes, is also famous for its application of electrically based processes in melting and remelting of high-alloy and high-temperature materials. These include the ESR (electroslag remelting) process, the ESC (electroslag casting) process, skull remelting based on electron-beam processes, plasma arc processes, and electric arc processes.

Along with the ESR process for ingot production is the commercial plasma arc remelt process for specialty steels, particularly where high nitrogen contents may be desired, as in austenitic stainless steels. Major industrial centers are now scattered throughout the USSR and are a major factor in high-alloy, high-strength, low- and high-temperature materials.

The ESR process was developed in response to the Western development of the VAR (vacuum arc remelting) process for producing very highly alloyed materials during the growth period of the jet engine age. The VAR and ESR processes utilize different purification and refinement processes that are extremely critical in very highly, complexly alloyed superalloys and high-speed tool steels. In water-cooled remelt systems, they also achieve relatively rapid (directional) solidification, minimizing segregation and coarse phase separation of undesirable impurity elements or elements that tend to form coarse brittle phases.

Over the years Dr. Medovar has been the prime mover in remelt processes and is credited for the ESR process, a major contributor to low-, intermediate-, and high-alloy materials. The developments include the production of hollow ingots for tube manufacture; fine grained, complex alloys that require relatively small amounts of hot work to produce useful, large structures; and the ESR *direct* remelting of discarded or scrapped rolls, heavy sections, axles, etc., without significant loss of expensive alloying elements, while maintaining a fine cast structure suitable for direct application or after a limited amount of hot work for shaping.

Recently (currently, in fact), the ESR process was modified to an ESC process, such that specific items or parts or shapes such as valve bodies, rolls, heavy structural sections, etc., can be remelted by the electroslag (ES) process in a ceramic crucible, under an appropriate highly basic lime slag, and directly cast into the same or similar shapes, or into still other shapes.

There can be no doubt that ES processes have been an extremely important part of the Soviet industrial effort for the production of very highly alloyed large (or small) shapes, with attractive as-cast structures and properties. The ESC process lends a flexibility in a very large country, where shipping of critical parts can be both expensive and slow; additionally, the reproduction of shapes of the damaged or scrapped parts offers very large rewards.

Dr. Boris Medovar has made all of this possible not only by the invention and development of ESR processes, but also by continuing developments, which include hollow ingots, shape castings including shape reproduction, and also a process that appears economical even for relatively simple alloy compositions.

Whereas the ESR process is relatively mature, and the ESC process is beginning to demonstrate its potential, there are numerous modifications to be expected for each process which will provide melting and casting refinements that will enhance the applicability and quality of steels and other metals and alloys for important, demanding applications in the industrial world.

Cambridge, Massachusetts                                        N.J. Grant

# Foreword

## The 30th Anniversary of Electroslag Remelting

The technique of electroslag remelting (ESR) is now 30 years old. At this juncture, we can try to sum up what has been achieved so far and to forecast what the next 10 to 15 years hold for ESR and also for all the other applications of electroslag processing, which are now collectively called *electroslag technology* (EST).

Let us first describe when ESR appeared and briefly dwell on the main stages of development of this process, which is the basis of specialized electrometallurgy (SEM), the most important branch of modern metallurgy.

The rapid development after World War II of new fields of engineering, such as aeronautical (and later aerospace), radiochemistry, and nuclear, posed new and unusual problems for metallurgists and materials scientists. Steel and other alloys were needed with hitherto unimagined combinations of properties, such as very high strength together with high plasticity and ductility. Metals were needed to construct equipment that had to have a given working lifetime in hostile environments. Tubes, bars, and sheets that would work under widely varying conditions—in hot hydrofluoric acid and boiling acid mixtures, in an ultrahigh vacuum, at cryogenic temperatures, under high pressure, at high temperatures, in neutron fluxes, and so on—were required.

It was not immediately obvious what these new materials should be, and it took time to come to the conclusion that the classical metallurgical methods current at that time (the 1940s and the 1950s) were inadequate. New approaches and technologies were needed for the production not only of carbon and alloy steels but also of many nonferrous metals and their alloys.

In the West, these needs were met by the widespread application of vacuum arc remelting (VAR). In the United States, where high-temperature steel and alloys were needed for the arms race, an enormous VAR capability was achieved in a relatively short time.

The Soviet Union could not follow this course, since its industry had not completely recovered from the war. Thus, it was necessary to find, quickly, other efficient but inexpensive, readily available ways to improve the economy and to strengthen defense. The solution was ESR. The first metals produced using this technology were high-temperature steels, acid- and oxidation-resistant steels, and high-temperature, nickel-based alloys. At about this time, there was interest in new metallurgical processes for the production of bearings, because the lifetime of aircraft engines (jet, turbojet, ducted-fan, and turboprop) was limited mainly by the lifetime of their roller bearings. The substitution of electroslag

(ES) steel for regular bearing steel made it possible to double engine lifetimes within one to one-and-a-half years of its introduction. A related new application of ES steel was in the production of parts for aircraft power trains. The results exceeded all expectations, with a doubling of the parts' lifetimes, thus opening up two new possibilities for improvement: either increasing engine lifetime or increasing engine speed. Both possibilities were exploited profitably. In a very short time, a large body of work was done in the investigation of ES metal quality. Now, more than 25 years later, the speed at which this small group of scientists and engineers worked seems unbelievable.

The construction of new production lines and specialized ESR departments was carried on simultaneously, with intensive research and testing of this new technology. The first monographs on the subject of ESR were published in the USSR in 1962 and 1963; they were soon translated and published abroad. Thirty years after its invention, ESR is still being used successfully in the Soviet Union.

Naturally, in the first stages of the development of ESR, the main emphasis was on high-volume production, since there was a shortage of ES steel in the Soviet Union; thus, many new ESR plants were constructed. By the middle of the 1970s, large ESR complexes were able to meet the ever-growing need for high-quality metal in all the various branches of industry. This metal was now widely used, and not only by its creators who laid the foundation of the whole trend 30 years ago. With continued growth in the metallurgical industry, the remelting of medium- and even low-alloy steel is now performed by ESR. This steel is used in blast-furnace jackets and all kinds of welded constructions and in tool steel and die steel production. Hundreds of steel and alloy quality metals are produced by ESR. It has become necessary to review what has been achieved and to assess ESR critically—not only its advantages but also drawbacks that have shown up in its new applications. This means rethinking the concept of ESR as a means of producing especially high-quality metal.

The last 15 to 20 years have been a time of radical change in ferrous metallurgy and in its subdivisions, such as specialized metallurgy (SM) and ESR. Modern ladle metallurgy (also known as secondary, injection, or furnaceless metallurgy) has radically changed the concepts of steel production. Thirty years ago, ESR was considered to be the only means of obtaining molten metal with extremely low concentrations of sulfur, oxygen, and nonmetal inclusions. Before the advent of ESR, there was no thought about the possibility of producing large quantities of so-called clean steel. With the introduction of injection metallurgy, the inexpensive production of millions of tons of low-alloy steel with total sulfur and phosphorus contents no greater than 0.02% became a reality. Moreover, vacuum-carbon deoxidation made possible the manufacture of steel with extremely low oxygen content of only several parts per million.

As a result of other advances in metallurgy, ESR was no longer regarded as the only method of obtaining extremely pure molten steel, and another feature of ESR became important. This was the sequential crystallization of a relatively small volume of molten steel in a fixed time interval. This type of crystallization gives a very high degree of physical and chemical homogeneity to the final product, whether it be a cast, forged, or rolled product.

It is evident now that a high-quality final product in steelmaking is available only if the process provides, first, high-purity molten steel, and second, appropriate control of the initial crystallization of the metal. The first problem has been successfully solved by the world's ferrous metallurgical industry; the second problem is still open. Of course, many important results were achieved by using continuous casting of the product instead of pouring into cast iron molds. There are also remarkable results in the preparation of large, steel ingots by vacuum-carbon deoxidation. The use of so-called internal molds (freezer molds) in casting is now also very successful—here we speak about what are now called *quasi-monolithic reinforced (QMR) steels*. Although much still remains to be learned about these new technologies, they already successfully compete with ESR in the manufacture of medium- and low-alloy steel.

Thus, 30 years after the invention of ESR, we are dealing once more with the same range of carbon and alloy steels that this whole technology started with. These are, first and foremost, alloy steels that are sensitive to the conditions of crystallization, for example, high-carbon steels (high-speed tool steels and ball-bearing steels), high-strength structural steels, and austenitic alloy steels.

In the development of ESR, priority was given to building up production capabilities. After this initial build-up phase, the quality of the ES steel and the efficiency of the process were emphasized. To obtain higher quality metal, it is necessary to use higher-purity starting materials, which contradicts the erroneous view that the remelting process is a way of correcting "mistakes" that occurred during smelting. Considerable research is being done in this area, from the initial smelting to the final phase of refining the ES metal.

The most important achievements of the modern metallurgical industry can be fully realized only by combining them with ESR. It is a mistake to compare furnaceless metallurgy to remelting metallurgy, but it is also a mistake to hold that converting to furnaceless metallurgy makes remelting metallurgy unnecessary.

It is well known that most of the ESR plants in the Soviet Union were built about 20 to 25 years ago. The majority of these plants still use outdated, first-generation ES furnaces that will have to be replaced by modern furnaces. In addition, such auxiliary equipment as power supplies and control systems need to be updated. It must be admitted that, in the last few years, we have lost the lead in the field of economic support and automatization of ESR technology. The quality of our steel is sometimes not as good as that of other countries; for example, some ESR technology indicators are better in the United States, Germany, and Japan.

For a variety of reasons, many new ideas that originated in our country were not widely used here but were instead realized more successfully abroad, for example, the liquid charge, alternating motion of the consumable electrode and freezer mold, bifilar winding of current leads to the ES furnace, ES piercing for the production of hollow castings, and the theory of automatic control (AC) of ESR. Another way that ESR technology became known abroad was through Soviet licenses.

It is impossible to discuss in detail in this review all the problems of ES-based specialized metallurgy in the Soviet Union, and so here we will mention the main ones: radical reconstruction of ESR plants, and refitting them with third-generation furnaces; reducing power consumption; and further improving the product yield and the ES metal quality.

In the last 10 years, remarkable results have been achieved in the theory and the technology of crucible ESR and ES casting (ESC). The production volume of metal in conventional ESR furnaces will not increase, since it has probably reached its optimum level. However, permanent-mold ESC and centrifugal ES casting will continue to grow rapidly.

A separate problem is that of increasing the productivity of ES furnaces. Sometimes, one still comes across statements that ESR is 30 years old but the process time is the same as it was in 1958, and that it should be possible to shorten this time by speeding up the charging of long freezer molds or by withdrawing castings from short freezer molds. It is maintained that this would increase the yield of ES metal per unit time. Such reasoning is incorrect, since there is an optimal melting time in ESR, and deviating from it would mean that the required metal purity and the physical properties and homogeneity of the ingot or casting would not be obtained.

Thus, what is the best way to increase the productivity of ESR? One of the solutions is to reduce the time spent on auxiliary operations and the idle time between melting runs. Single-post furnaces should be replaced by multipost ones, and auxiliary operations should be mechanized. There is great potential for increasing ES metal production by simply remelting longer electrodes. This will help to reduce the amount of scrap from crop ends and will increase the ingot-to-product yield. We already have at our disposal ESR furnaces capable of producing ingots one meter in diameter and 6 meters long.

One of the most effective ways of increasing the productivity of ESR furnaces is the correct application of automatic control systems (ACS), but the decision to use ACSs should be made carefully. An ACS is advantageous only when it can be used with the appropriate equipment. That means that ACS work with ESR equipment operating under well-defined, reproducible conditions, such as secondary supply voltage, feed rate of the consumable electrode, or relative travel speed of the crystallizer and the ingot.

As is well known, ESR is characterized by a high thermal inertia. The temperature distribution in the molten slag layer changes slowly, and the rate of change is obviously lower for larger baths. Because of this high inertia, a fast-reacting ACS is not required for ESR. At present, many countries are experienced in using ACS for ESR processing. It should be mentioned that, in recent years, the ACS in use abroad had advantages over those used in the Soviet Union. We do not mean physical or modeling advantages, but rather better instrumentation. As an example, the development of high-quality sensors for measuring the weight of the consumable electrode and the ingot has helped some foreign companies to provide highly effective ACS.

ESR is very power intensive. Usually, specific power consumption is about 1,500 kilowatt-hours (kW-h) per ton of metal produced. With the use of a bifilar circuit, the power consumption is 20% to 25% lower. Water cooling in the melting zone does not appreciably reduce power consumption, but it is possible to reduce it by crucible ES melting, which is a kind of slag-autocrucible melting. In this method, specific power consumption is closer to the theoretical power consumption and is comparable to that obtained in electric arc or radiofrequency-induction melting, or almost two times lower than that for conventional ESR. However, for modern arc melting, the specific power consumption is no higher than 500 kW-h/ton. Of course, it is not always possible to use crucible remelting, and therefore, one should not reject other, simpler ways of reducing power consumption, such as increasing the electrical conductivity of the molten slag and reducing the thickness of the slag layer later. In the Soviet Union, it has been traditional to use thick slag layers, since in shallower baths, electrode depth must be controlled more accurately. In comparison, the molten slag layers for ESR in foreign countries are almost one and a half to two times shallower than in the Soviet Union.

In considering the overall economics of the ESR process, we should again mention that the prime cost of ES metal is determined by shop expenses. In the Soviet Union, ESR plants are usually overstaffed, sometimes on purpose. Furthermore, it is essential to reduce the cost of the slag, and it is also reasonable to save on auxiliary operations such as assembly and the arrangement of electrodes.

As was already mentioned, the development of ES technology in the last 10 years has been characterized by the appearance of permanent-mold ESC and centrifugal ESC. Currently, ESR is most popular in ferrous metallurgy, although all types of ESC are used in manufacturing. The combination of crucible remelting with different, more modern mold preparation is a very promising development in the area of ESC.

What is the outlook for the development of ES technology in the next 10 to 15 years? A substantial increase in production of ESR steels is not expected. Of course, there will be some increase through the upgrading of existing plants. But more significant growth of ESC is expected in ESR plants that work in combination with industry. Moreover, it is more sensible to increase capabilities for joint use by several branches of industry than to provide each factory with its own ES shop. It is much easier to build large regional factories with all the necessary steelmaking and heat-treating equipment. The modern crucible ESR and ESC unit is self-contained and can be used any place electric power is available. The power consumption of this unit is not high, and it can be used as a movable unit for different applications.

In recent years, the problem of reusing scrap from metalworking has arisen. The conventional way of recycling this scrap was to crush it and melt it by hard charging a furnace. This is efficient and is still used for ordinary, inexpensive carbon steel and cast iron waste, but it is inappropriate for alloy steels or even low-alloy steel scrap. However, it is unreasonable to throw away such valuable metals as tungsten, niobium, and molybdenum, which is essentially what happens when unsorted scrap is remelted together.

The problem of recycling alloy steel scrap can be handled by ES-crucible remelting. Scrap can and should be utilized at the place it is produced. This scrap
may come from various sources, such as obsolete metal-cutting or forming or
stamping machinery; it can also contain nonferrous metals, for example, copper
and copper alloys. We have had enough experience with ESR to regard it as a
viable method of recycling valuable metal scrap. In this respect, the experience
of the Ukrogossnab is interesting. The first ES shop in the Soviet Union for
recycling alloy steel and bronze scrap was built in the Kiev regional service center. Here scrap is shaped into various forms such as cubes, slabs, or bars.

Traditionally in manufacturing, ESC plants use either commercial bar stock or
their own forgings for consumable electrodes. This means that thousands of tons
of high-quality, formed metal are remelted to produce castings, which then end
up as forgings or billets similar to the electrodes. Thus, it is evident why it is
more efficient, first, to use cast consumable electrodes and, second, to develop
ESC technologies that are based on the use of molten refined metal—in other
words, technologies that do not involve remelting as such but rather depend on
pouring the molten metal and synthetic slag (both of which are available in
ordinary steel mills) into the mold together. Nowadays in industry, rather small
consumable electrodes are used: They are 60 to 150 mm in diameter and up to a
few meters in length. It is impossible to cast such electrodes by using special
molds, as when large consumable electrodes 600 to 1,200 mm in diameter are
made.

There is now a new way of casting long, relatively thin consumable electrodes
due to the recent appearance of a new generation of horizontal, low-pressure
machines (HLPM) for continuous casting. The first HLPMs were built by
the Kharkov Institute of Ferrous Metallurgy (UkrNIImet). Currently, such
machines are operating in manufacturing in the Soviet Union, but they are
used only for cast iron, as in the "Centrolit" factory in Kaunas, or for nonferrous metal casting. These machines are not yet available for steel casting.

Abroad, HLPMs are widely used. As of 1985 there were over 30 machines,
and there are many more now. These machines have made it possible to solve a
number of problems. One of these is providing ES machines with cheap, continuously cast, consumable electrodes (here we mean all kinds of ES equipment
that use consumable electrodes, for example, ESC furnaces, equipment for
permanent mold and centrifugal ESC, and ESR furnaces). Another problem is
providing steel mills with cheap, continuously cast, ES ingots for further use in
either as-poured form or for hot or cold forging.

All of the above holds, provided the manufacturing plant has its own steelmaking facilities. If the plant has, it is essential to situate the HLPM in the
vicinity of a steelmaking furnace or steel foundry. If, however, the plant does
not have its own molten metal, then it has to rely on induction furnace melting
of scrap. This scrap can come from various sources within the plant, such as
obsolete machinery.

This book is a collection of papers devoted to the 30th anniversary of ESR and
differs in several respects from the two previous collections that were published
on the 20th and 25th anniversaries of ESR. First, it reflects the wider geographi-

cal distribution of EST plants; second, we have included several papers that compare QMR steels with ES-refined steels. In addition, there are numerous papers from foreign countries describing the status of ESR there.

We hope that this collection will be useful to those using ESR and that it will help to advance the field of specialized metallurgy and manufacturing.

Kiev, Ukraine, USSR                                                B.I. Medovar
                                                                   G.A. Boyko

# Contents

# Abbreviations

ACS     Automatic control systems

ARS     Automatic regulation system

AW     Arc welding

CCM     Continuous casting machine

CESC     Centrifugal electroslag casting

EBM     Electron-beam melting

ESC     Electroslag casting

ESR     Electroslag remelting

EST     Electroslag technology

ESW     Electroslag welding

IM     (RF) Induction melting

MHTP     Multistage high-temperature processing

NMI     Nonmetallic inclusions

OAM     Open arc melting

PESC     Proportional electroslag casting

PM     Powder metallurgy

PMESC     Permanent-mold electroslag casting

QMR     Quasi-monolithic reinforced steel

REM     Rare earth metal

ESRP     Electroslag remelting under pressure

SCCM     Semicontinuous casting machine

SEM     Specialized electrometallurgy

TP     Technological process

UFM     Utilization factor of metal

UI     Ultrasonic inspection

VAR     Vacuum arc remelting

PART 1

# Electroslag Technology in the Soviet Union

# 1 Contemporary Electroslag Crucible Melting and Casting, and Its Future Outlook

B.E. Paton, B.I. Medovar, G.S. Marinski,
V.L. Shevtsov, and U.V. Orlovski

The method of electroslag (ES) remelting of the consumable electrode in a water-cooled crystallizer was developed at the E.O. Paton Electric Welding Institute 30 years ago and is now well known and widely used all over the world. Electroslag remelting (ESR) and electroslag casting (ESC) are used in the production of high-quality ingots and castings comparable to forgings in their physical and mechanical properties. The castings are used for cement kiln bands, crankshafts, high-pressure vessels, stamping forms, cutting tools, various types of hollow semifinished products, and so on.

Both ESR and ESC are single-stage methods. They provide simultaneous melting and solidifying of metal inside a water-cooled crystallizer. The advantages of this feature are that there is no contact between the refined molten metal and the atmosphere and the melting unit and the casting mold; also, the volume of the molten metal bath is minimized. There are, however, drawbacks to the single-stage process, for example, a limitation on the variety of casting shapes. Casting usually, are simple in form and have a small length-to-diameter ratio. Another drawback is that the complex water-cooled copper crystallizers that are used are expensive and difficult to manufacture and that the consumable electrodes are complex in shape, which limits the use of ESC for remelting used and damaged parts and obsolete machinery. A two-stage process makes it possible to eliminate some of these drawbacks. In the first stage, ESR is used to produce the molten metal in the crucible furnace; in the second stage, the metal and slag are poured together into a mold. The realization of this idea in the E.O. Paton Institute permitted the formulation of two new production methods: centrofugal electroslag casting (CESC) [4] and permanent-mold electroslag casting (PMESC) [2]. The main difference between these methods is that, in the case of CESC, the casting mold rotates, whereas, in PMESC, the casting mold does not rotate. These methods are rather simple, economical ways to produce castings of various shapes with appropriate properties for subsequent forgings.

In contradistinction to ESR and ESC, metal melting during CESC and PMESC takes place in a crucible with refractory linings, which are in contact with the molten slag at temperatures of about 2,000°C.

Benefits of the new production processes are noted in detail in [1 to 3]. We list only the main items here. First, it is possible to use consumable electrodes of worn-out or obsolete tools or forging and roll crops and to remelt both whole

Authors' affiliations: E.O. Paton Electric Welding Institute, Academy of Sciences, UkSSR.

(a)          (b)          (c)          (d)

**Fig. 1.1.** Castings of various shapes produced by CESC (a), (c), (d) and PMESC (b).

pieces of scrap and chips, and to assemble all the necessary equipment in a single shop due to its small size. Also, the high-quality ingots of various shapes already closely resemble the final product; ES crucible furnaces are highly efficient; and the simple equipment is inexpensive.

Since the end of the 1970s, when the first experiments on CESC and PMESC were performed, both methods have been widely used in industry. Now many plants in the Soviet Union have CESC and PMESC departments that produce semifinished products for couplings for high-pressure pipelines, dies, bars, gears, half-couplings, sizings for tube mills, and so on (Fig. 1.1). These products are made of carbon and alloy steels, tool steels, die steels, high-speed and bearing steels, stainless steels, high-temperature steels and alloys, pig iron, and copper and its alloys, that is, practically all types of steel and nonferrous alloys currently used in industry. Numerous tests and analyses peformed on castings made by CESC and PMESC—and on finished parts and tools [1–3]—show that this metal has mechanical properties that are not inferior to those of forgings made of the same type alloy. In the E.O. Paton Institute, new equipment has been designed to obtain preforms in a wide range of weights, from several kilograms to several tons. The most popular equipment is mass-produced. It must be mentioned that conventional equipment for CESC and PMESC is so simple that it can be made in any ordinary manufacturing plant.

Licenses for CESC and PMESC have been sold and are now successfully being used in the United States, India, Bulgaria, and other countries. Of course, such new technologies cannot be universally used and are meant to be used only in appropriate fields. The weight of CESC and PMESC castings varies from several kilograms to several tons, and the castings can be made in a variety of shapes. The types of castings represented in Fig. 1.1 give a general idea of the possibilities of the new process. The question of whether a desired product can be manufactured can be answered only after thorough analyses and tests. However, CESC and PMESC cannot completely replace modern methods of precision casting. Depending on the weight and the batch size of the castings, one of the several versions of CESC or PMSC should be chosen. If the casting weighs more than abotu 50 to 70 kg and several hundred to several thousand units per year are produced, then the conventional version of CESC or PMESC [1, 2] is used. This means that there is only one casting per melting run; that is, all the metal held in the crucible is poured into the casting mold with the slag. In the case of lighter preforms and smaller batch sizes, another version should be chosen. As the volume of metal held in the ES crucible furnace increases, productivity also increases. Thus, increasing the volume of the crucible and dividing the molten metal into several portions to approximate the weight of the casting make it possible to increase the number of castings per melting run. This division can be accomplished by using multisectional molds or carousel-type centrifugal castings machines. It is possible to use a continuous process by continuous electroslag ES melting of the starting metal, and by tapping the molten metal when a sufficient quantity has accumulated in the crucible.

Depending on the application, CESC and PMESC castings can be divided into two main groups: castings for stamping dies or other specialized tools or repair parts and castings for use in manufacturing. Castings from the first group can include, for example, preforms for sizings for tube mills, die inserts, shaping mill rolls, web and tread rolls, and large-scale cutting tools. Examples of preforms for repair parts are preforms for casting crane wheels (Fig. 1.2), bulldozer rollers (Fig. 1.3), and ore mill linings (Fig. 1.4).

Usually, CESC and PMESC castings are used to satisfy the internal needs of the factory where they are produced; in that case, the starting material comes from worn-out or rejected tools. When conventional production methods are being used, this kind of initial material is regarded as scrap and is remelted together with other ferrous scrap in open-hearth furnaces. When this is done, expensive alloy tool steels or structural steels, such as 5XHM, 60C2XFA, and 40X2HMA, are degraded, since during melting they are mixed with other steels, the alloying elements are diluted or burned out, and the metal has to be alloyed again. In contrast, remelting the melt in an ES furnace barely changes its

**Fig. 1.2.** Preform of a crane wheel produced by CESC.

**Fig. 1.3.** Preform of a roller for a heavy bulldozer produced by CESC.

chemical composition [1, 2]. Castings are obtained by pouring metal into a metal mold or a centrifugal castings machine. After that, the casting must be worked before it can be used. Moreover, the shape of CESC or PMESC castings is close to the shape of the final product so that very little working is necessary. As soon as tools or parts are worn out, the whole cycle can be repeated.

We can remark that, for remelting in the crucible furnace, it is not necessary to fabricate the starting material into consumable electrodes as is required for ESR in a water-cooled crystallizer.

**Fig. 1.4.** Ore mill lining produced by PMESC.

Introduction of the above closed cycle of tool production from internal scrap provides a remarkable flexibility in manufacturing and repair services, eliminates dependence on external metal suppliers, reduces expenses for transportation, and so on. For example, the experience at the Nikopol South-Pipe Plant shows [2] that the prime cost of dies for tube drawing produced by PMESC is one half that of forged dies and that the working life of these dies is the same or even higher. Similar results were obtained in other plants.

Preforms for tools and parts can be cast not only by remelting waste but also by using metal from billet crop ends or off-grade billets. It is also possible to add alloying materials to the starting metal to obtain castings with a structure that is not available in forgings or in rolled billets. For example, castings for web and tread rolls from steel 150XHM could be produced by CESC when consumable electrodes made of 40XHM steel were used.

At the E.O. Paton Institute, CESC and PMESC technologies have been devised to produce castings from high-speed cutting steels. Test results on these castings [3] made it clear that cast tools made of metal produced by ES crucible melting can be widely used in industry. Thus, it would be reasonable to use worn-out and obsolete tools for starting materials. But because it is inconvenient to use this material to obtain consumable electrodes by welding separate pieces together, it is more acceptable to melt down the lump material along with the permanent electrode. Remelting of consumable electrodes made of metal melted in an induction or arc furnace is also possible.

In addition to producing steel castings, CESC and PMESC can be used to make castings of copper and copper alloys, for example, parts for water-cooled crystallizers in continuous casting machines, copper crystallizers, and ES furnace bottom plates. For this purpose, the worn-out and obsolete tools are remelted as the consumable electrode, and copper lumps or chips are remelted using a permanent electrode. The process is performed in a crucible with a carbon lining, which significantly reduces the oxygen content of the alloy produced. It is then possible to obtain very dense cast metal parts that do not require further shaping.

These two methods provide a number of opportunities for the production of cast preforms in manufacturing. The use of CESC and PMESC is advantageous in the following cases: When CESC and PMESC castings can be substituted for forgings or other high-quality preforms by providing more efficient use of metal and reducing casting time; when it is necessary to obtain castings of extremly high quality; when it is impossible or unprofitable to produce the preforms by any other method; and when a large quantity of scrap is available for ESR in the crucible furnace to obtain castings of the required quality.

In the production of parts for manufacturing, PMESC and CESC are used to make sun gears for large quarry dump trucks, carriers for large transporter trucks, diamond drills, different types of fittings, and so on. In each case, CESC and PMESC castings replaced forgings, which notably increased utilization factor of metal (UFM) and reduced machining time and chip waste (Fig. 1.5). In the production of CESC and PMESC castings for manufacturing, it is not always convenient to use machine scrap for raw materials, as described in the previous

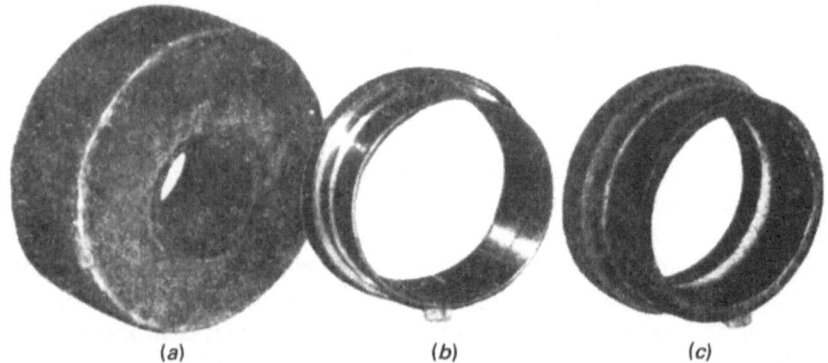

(a)                          (b)                          (c)

**Fig. 1.5.** CESC castings (c), the finished product (b), and the forging from which the part used to be made (a).

example. It is reasonable, nevertheless, to use cheap raw materials such as forging and plate steel crop ends, billet waste, and nonstandard billets. To illustrate the procedure, we describe the production by CESC of T-joint preforms at the Fastovski Gas–Crude Oil and Gas Using Equipment Plant. Here the initial material consists of plate steel crop ends (20 to 30 mm thick) of 15XCHD steel, which are produced at another plant, to fabricate by any other method. In another case, billet trimmings are used as raw material to produce CESC castings for large diamond drills [7].

The cost of CESC and PMESC is mainly determined by the cost of the re-melted metal and the complexity of the casting shape, with the average being 300 to 800 rubles per ton casting weight. To make a rough estimate of productivity, one can consider that the yearly productivity in tons $(q)$ of an ES furnance working two shifts per day is numerically equal to twice the capacity of a crucible in kilograms $(v)$ $(q = 2v)$, that is, if the crucible holds about 200 kg of molten metal, then its yearly productivity is about 400 tons. If the weight of the casting is 200 kg, this mean that 2,000 castings are produced per year.[1] Of course, when the molten metal is poured into multiple molds, the number of castings will be different. These estimates can be helpful in determining the number of units to be installed in new CESC and PMESC shops [8]. Sometimes a plant that needs ES preforms for parts and tools is not able to produce them due to a lack of space or resources. The solution is to organize regional centers that work on consignment for separate plants. An example of this kind of ES center is operating now in the "Metallomashprom" company in Kalinovka, in the Kiev region.

Now we point out the main directions of the further development of ES crucible melting and casting.

1. Improving CESC and PMESC flows. This can be accomplished, first, through the introduction of new ACS to provide optimal working parameters and repeatable melting results. Second, the development of new slag composi-

---

[1] This calculation should be regarded as an estimate [editor's note].

tions and various methods to alter the melt in the crucible will lead to further improvements in the quality of the cast metal, permit the production of castings with more complicated shapes, and make possible the use of casting mold combinations containing, for example, ceramic or sand-clay parts.

2. Further increasing the capacity of melting crucibles in order to increase the weight of single castings and to allow the use of simultaneous or consecutive pouring into several molds.

3. Developing new technologies to obtain more castings (up to 20 or 30) per molten metal charge.

4. Working out the technology of CESC and PMESC with the introduction of internal reinforcing elements in the castings to produce composite castings with special properties. Also very promising are processes whereby macrocoolers are introduced into the mold or into the molten metal stream, thus eliminating metal overheating and improving the structure of the casting.

5. Further developing ES crucible melting with permanent electrodes for remelting lump scrap and chips of different alloys.

6. Using ES crucible melting in combination with other modern casting methods, such as die casting, to broaden the range of castings.

7. Further perfecting equipment for CESC and PMESC; creating a new generation of centrifugal machines and other specialized equipment; and constructing high-efficiency, completely automated CESC and PMESC devices.

8. Developing ES crucible melting technology using a molten charge.

These new developments in CESC and PMESC give only a general idea of their potential. Many new developments are occurring in the course of current work.

## References

1  B.I. Medovar, G.S. Marinski and V.L., Shevtsov, "Centrifugal electroslag casting," Znanie Kiev, USSR, p. 48 (1983)

2  B.E. Paton, B.I. Medovar, U.V. Orlovski, "Permanent mold electroslag casting," Znanie Kiev, USSR, p. 64 (1982)

3  B.E. Paton, B.I. Medovar, and G.S. Marinski, "Electroslag crucible melting and new casting technologies developed on its basis," Proc. 8th Intl. Conf. Vacuum Metallurgy, Lintz, Austria, pp. 1335–1369 (1985).

4  N.V. Zhuk, B.I. Medovar, et al., "Analysis of 1H16N4B steel preforms produced by centrifugal electroslag casting," Problems of Specialized Metallurgy, 2, pp. 14–19 (1987).

5  G.S. Marinski and A.P. Zinkovich, "Production of preforms form high-speed R9 and R6M5 steel by means of centrifugal electroslag casting," Report of E.O. Paton Institute, No. 22 (1986)

6  G.S. Marinski, A.V. Chernets, and A.P. Zinkovich, "Production of drill bits by centrifugal electroslag casting," report of E.O. Paton Institute, No. 23, p. 1 (1986)

7  V.V. Panasyuk, K.B. Katsov, and T.A. Ilyk, "Production of drill bits from Electroslag Steels," Problems of Specialized Electrometallurgy 1 (1988)

8  "Equipment for electroslag remelting and casting," catalog-reference, Naukova dumka. Kiev, p. 32 (1986)

# 2 Electroslag Technology as a Means of Improving the Design and Properties of Parts Used in Corrosive Environments

V.V. Panasyuk, K.B. Katsov, V.I. Kovalenko,
V.P. Rudenko, and A.B. Kuslitski

The technique of electroslag remelting (ESR), which was invented at the E.O. Paton Electric Welding Institute 30 years ago, is still a widely used method of metal refining. The widespread use of ESR in industry is based on the results of numerous tests and analyses of electroslag (ES) product quality in comparison to alloys obtained by conventional remelting, other refining methods, and particularly vacuum arc remelting (VAR). As a rule, the basic physical and mechanical properties of ES products are comparable to those of VAR alloy, and some are even better.

Initially, the quality of ES alloy was judged on the basis of conventional mechanical tests, which did not demonstrate fully the advantages of the new technology. Although these tests showed, at most, several tens of percent improvements for ES alloys, the results of commercial operation of parts made of ES alloy showed increases in lifetimes of several hundred percent. Machine parts and construction elements made of ES alloys were more reliable and lasted longer due to their improved ability to carry loads.

That fact was proved by the first fatigue tests of steel ShH15 obtained by ESR [1 and others]. From the middle of the 1960s, such tests, which were frequently performed, were conducted on various ES steels used in different industries. As a result, the many advantages of ES metal became evident. Nevertheless, in some publications, either the fact that nonmetallic inclusions affect the fatigue limit [2] was rejected or the necessity of metal refining was disputed because the increase in production costs did not justify the improvement in mechanical properties. Authors in other publications [3] asserted that refining is advantageous only in two cases: to obtain very high tensile strength steels ($\sigma_{hts} > 1,400$ MPa) in massive sections and to obtain medium- and high-strength steels in thin sections (strip, wire, etc.). Those erroneous views were based on incomplete information, since the authors usually did not take into account the real working conditions under which the finished parts and constructions would be used, the physical and mechanical properties of the material, and the possibility of a corrosive working environment.

Thus, a reliable analytical method was used to investigate the capabilities of ES metal, one that would take into account real working conditions. This

Authors' affiliations: V.V. Panasyuk, K.B. Katsov, V.I. Kovalenko, and V.P. Rudenko, G.V. Karpenko Physico-Mechanical Institute, Academy of Sciences, UkSSR; A.B. Kuslitski, Iv. Fedorov Printing Institute, UkSSR.

problem could not have been solved completely by metallurgists, chemists, or materials scientists. The solution was to be found in the development of a new scientific discipline—the physicochemical mechanics of materials, which is being pursued successfully in the G.V. Karpenko Physico-Mechanical Institute (FMI).

The author of the electrochemical adsorption theory of corrosion fatigue, G.V. Karpenko, [4] also suggested new models for adsorption fatigue [5] and hydrogen fatigue [6]. It was found that numerous factors affect the resistance of steel to fatigue and that this depends on the nature of the environment and on the physicomechanical properties of the metal, especially of its surface layers [7].

These new approaches made it possible to test and analyze ES metal under conditions close to actual working conditions and clearly showed the advantages of ES metal.

In 1962, Karpenko discovered the favorable effect of ESR on corrosion and corrosion fatigue resistance of steels [4]. He determined that ESR increases the corrosion resistance in a 3% solution of sodium chloride by 15%, and the corrosion fatigue limit by 20% after normalizing and by 40% after martempering. Thus, he showed the possibility of substantially increasing resistance to corrosion fatigue by high-strength steels.

The results of the joint research carried out by scientists from the E.O. Paton and G.V. Karpenko institutes showed that ESR notably reduces the sensitivity of steel to the size effect during fatigue testing [8]. Tests on ES and open-hearth steel 12XH3A revealed (Fig. 2.1) that the size effect varies from case to case. For samples 5 mm in diameter made of open-hearth and ES steel, the difference in tensile strength was only 20 MPa, but for samples with a diameter of 25 mm, the difference was 60 MPa. As a result, it was concluded that not only was it

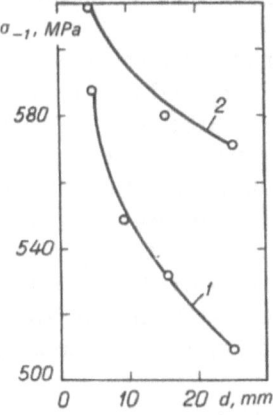

**Fig. 2.1.** The influence of a test sample of diameter $d$ on the tensile strength of steel 12HN3A obtained by two melting methods [8]: (1) open-hearth melting; (2) ESR.

advantageous to use ESR to produce large-scale parts but also that it is unreasonable to perform tests on small-diameter samples to predict the performance of large finished products. Moreover, it was shown [9] that for products
made for ES metal, it was necessary to allow for the so-called technological
*heredity*, that is, in order to obtain full benefits from the refined metal, high-
quality machining was necessary.

The favorable effect of ESR on steel properties depends to a large extent on
the eliminating and dispersion of nonmetallic inclusions, the harmful effects of
which show up under complex stresses and large strains. It was shown that for
those types of loading requiring a high-quality surface finish, the maximum effectiveness of refined steel is revealed by the following tests: low cycle bending
fatigue, cyclic contact loading, and cyclic twisting [10].

Special attention was paid to an investigation of ES metal service in corrosive
environments [10–13 and others]. It was found that refining substantially increases local and general corrosion resistance of carbon and alloy steels in various corrosive environments at room temperature and elevated temperatures.
Moreover, the corrosion rate of refined steel is usually several times lower than
that of conventionally melted steel.

The advantages of ES alloys are especially noticeable in the case of the combined effect of corrosion and mechanical stress (stress corrosion cracking; corrosion fatigue). During corrosion cracking test, boiler steel 16GHMA was held in a
30% solution of sodium hydroxide under the following conditions; temperature,
100 °C; deformation rate, 0.001 mm/min; and stress, 38 to 40 MPa. Samples in
the transverse direction from open-hearth steel cracked in 260 hours, whereas
similar ES alloy samples under a stress of 48 MPa, did not crack even after
1,800 hours. The increased lifetime of the ES alloy in the alkaline electrolyte,
compared to that of the open-hearth metal, is based on its high resistance to
pitting and to corrosion cracking [13].

Testing for low-cycle corrosion fatigue resistance of ES metal [10, 11, 14]
shows that refining notably increases the lifetime of transverse and especially
longitudinal samples [10, 11].

Refining reduces the intensity of local anodic dissolution of the metal, as
shown by the reduction in the spread of the local electrode potential on the
surface of low-alloy steel after ESR to one half that of ordinary steel [14]. The
increase in the microscopic electrochemical homogeneity of refined steel reduces
anodic current and static cell potential. In addition, the magnitude of the electrode static potential depends on inclusion content, which means that metallurgical *heredity* affects the electrochemical properties of the steel [14].

It was important not only to show the superiority of ES metal but also to
analyze it and explain the reasons behind it. In this respect, it was useful to
consult the available theoretical and experimental data on the role of nonmetallic inclusions in steel failure.

It was determined that nonmetallic inclusions are stress concentrators [3, 15]
and the source of significant thermal stresses [10, 16, 17]. They also cause sub-
structural changes in the metal matrix [16, 17] and affect dislocation movement
and the mechanism of steel failure [18]. This is why nonmetallic inclusions significantly affect both steel ductility and crack initiation [3, 8–14, 16, 17, 19, 20], as

well as stress at the crack tip, its development rate [16, 21], and suseptibility to brittle failure [19, 20, 22]. Nonmetallic inclusions have a particularly detrimental effect on the fatigue behavior of steels because fatigue cracks develop mainly at nonmetallic inclusions [3, 10].

The shape and size of nonmetallic inclusions are important parameters that influence the fatigue life of metals. Usually, reduction of inclusion size increases fatigue resistance; various data put the critical size of nonmetallic inclusions in the range between 3 and 35 $\mu$m. A similar effect is observed for spheroidization of inclusions.

In corrosive environments, nonmetallic inclusions display some new properties. The thermal tensile residual stress and stress concentration in the vicinity of an inclusion affects the interaction of the metal with aggressive liquid media [10–14]. The hydrostatic stress that develops around inclusions causes the corroding species from the liquid medium to diffuse into the metal. Due to this diffusion, the contact area between the corroding species and metal increases which speeds up adsorptive, corrosive, and other interactions between the metal and the liquid medium. It should be noted that boundary between the metal and the inclusion serves as a concentration point for dislocations, vacancies, impurities and other defects that migrate from the matrix, especially under the influence of deformation. It is evident that these boundaries are the most chemically active regions in deformed metals. In surface-active media, nonmetallic inclusions probably intensify adsorptive effects.

The method of analysis of the micro-electrochemical homogeneity of metals, developed by Karpenko et al. [23–25], has made it possible to study the most important aspects of the behavior of nonmetallic inclusions in corrosive media. During tests with intentionally contaminated steel [20], it was found that, in corrosive environments, nonmetallic inclusions act as cathodes with respect to the metallic matrix and are arranged in order of decreasing cell potential, that is, sulfides, alumina, silica, and titanium nitrides (TiN). Interestingly, the same order is followed for increasing electrical resistivity. The distribution of potentials in the vicinity of inclusions constitutes a significant potential gradient on the matrix–inclusion boundary, which causes a high-current density in this region [11]. Indeed, microscopic observations show that inclusions float to the surface in a short period of time, due to the fast dissolution (etching) of the metallic matrix at the inclusion boundary. This effect is strongest for sulfide inclusions and weakest for TiN inclusions, which corresponds to the ratio of potential gradients.

The effect of nonmetallic inclusions on the magnitude and distribution of cell potentials can be explained by differences in their semiconducting properties. This fact was confirmed in other studies, e.g., see [26]. This work describes a study of dissolution topography of steel 03H17N14S2, in which the nonemtallic inclusions are complex oxysulfides, 1 to 18 $\mu$m large. Under anodic activation in a neutral chloride solution (5 $N$ NaCl; 30 °C), pitting was initiated only at inclusions. Dissolution occurred initially at localized areas on the perimeter, and subsequently over the entire perimeter of the inclusion, to form narrow grooves, which grew into pits.

Microscopically, long-term (9 to 11 days) contact of samples with naturally aerated 0.1 $N$ NaCl and sodium iodide (NaI) solutions caused red corrosion rings to develop around some of the nonmetallic inclusions. By X-ray microanalysis it was determined that the rings contained iron and oxygen, but not the elements characteristic of inclusions [26].

The effect of mechanical loading intensifies corrosion, and this holds true for all types of inclusions. Recently, it has been determined [27] that in a corrosive environment the effect on failure kinetics of electrochemical conditions at the crack tip is as important as the effect of the stress concentration factor. This is the case for both static loading and cyclic loading. Thus, nonmetallic inclusions in as-yet unfailed metal, can, by changing the local cell potential, significantly increase corrosion.

In hydrogen-rich environments, nonmetallic inclusions collect hydrogen [6]. The most devastating inclusions in this respect are those with a large surface area (e.g., needle-like sulfides, soft silicates, and needle-like, brittle alumino-silicates), which are the sources of failures in hydrogen-rich environment [10, 11]. These data showed the significant effect of ESR on the increase in the service life of carbon and alloy steels through the elimination and dispersion of nonmetallic inclusions, especially sulfides.

Electroslag remelting affects steel characteristics in the following ways: It increases electrochemical homogeneity and stress corrosion resistance; it eliminates the linear dependence of the static cell potential on the inclusion content; and it correlates with the density of the steel. These effects demonstrate that the general improvement in steel properties by ESR is determined not only by refining but also by crystallization effects of the water-cooled crystallizer, as a result of which one obtains castings containing axially oriented, columnar crystallites with less segregation of impurities and increased plasticity.

Due to their high density and homogeneity, ES metal castings can be successfully used in products for service in corrosive environments. The corrosion resistance and fatigue limit of these alloys is higher than those of hot-worked materials obtained by usual melting methods [12, 28]. The efficient use of the properties of ESR cast parts can remarkably increase the fatigue limit of construction elements, for example, in the production of structural shapes [29]. The replacement of a welded product with a thickness of 40 mm by an ES casting increases the tensile strength more than 1.5 times because the concentration of stress points is reduced (Fig. 2.2).

To take full advantage of the material properties of ES, the steel should be heat treated, for example, by a cyclic heat treatment. It was determined that repeated heat treatments, which reduce austenitic grain size and work hardening, notably increase the corrosion resistance and yield stress of 40H steel in air and in a hydrogen-rich environment [28, 30].

The use of centrifugal electroslag casting (CESC) provides new opportunities in the production intricately of shaped parts. The properties of the cast metal, which has been obtained by CESC at an optimal mold rotational speed, are, after appropriate heat treatment, better than those of metal shaped from open-hearth steel and very close to those of ES forgings (Fig. 2.3). Tests in corrosive environments had analogous results.

The use of rare earth metals (REM), either as separate alloying elements or as

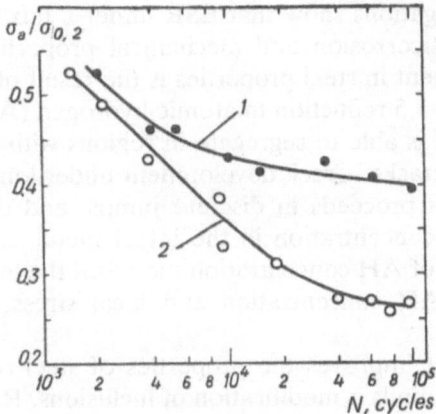

**Fig. 2.2.** Fatigue strength of metal stamps fabricated in one piece by ES casting (1); welded from sheet steel (2).

oxide fluoride compounds, which are added into the flux, has a high potential for increasing ES steel service in corrosive environments [31].

One report [31] analyzes the influence of fluoride- and oxide-based fluxes on the physical and mechanical properties and hydrogen contents of low-alloy ES steels. The effectiveness of using fluxes based on REM compounds was based on the results of conventional mechanical tests, such as conventional fatigue tests, low-cycle fatigue tests, and corrosion resistance and stress corrosion tests. The results of a bending-rotational fatigue test ($5 \times 10^8$ cycles) showed that remelting under an REM-based flux increases the tensile strength of smooth samples by 40%. Remelting under .a flux based on REM compounds also increases resistance to corrosion cracking so that the $K_{scc}$ of this steel is higher by 20% than that of steel remelted under an ordinary flux [31]. This low-cycle fatigue behavior under various deformations and in various environments also demonstrated the advantages of metal melted under a REM-based flux.

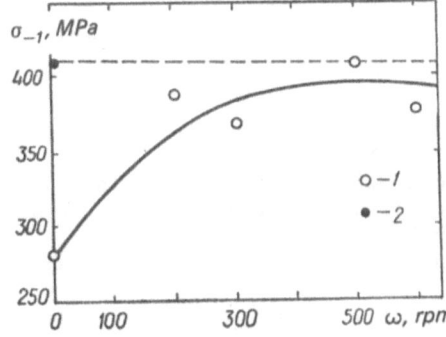

**Fig. 2.3.** Dependence of the endurance limit of centrifugal, ES cast, low-alloy steel on the rotational speed of the mold: (1) Centrifugal ES; (2) ES forging.

Thus, these investigations show that ESR under a flux based on REM compounds improves anticorrosion and mechanical properties of cast, low-alloy steels. The improvement in steel properties is the result of several factors. One of these factors is the 1.5 reduction in atomic hydrogen (AH) content. It is well known that hydrogen is able to segregate in regions with increased local stress and there to initiate cracks. Crack development under long-term loading in the presence of hydrogen proceeds in discrete jumps, and the incubation period depends on the AH concentration in the intact metal as well as on the local stress. The reduction of AH concentration increases the level at which the critical combination of AH concentration and local stress, necessary for crack propagation, occurs.

Another factor that improves the properties of steel remelted under a flux based on REM compounds is modification of inclusions. Rare earth metals have been seen to be incorporated in ES steel in nonmetallic inclusions. Metallographic examination shows that remelting under an REM-based flux changes the composition, shape, size, and quantity of nonmetallic inclusions. For steels remelted under ANF-6, the main inclusions are angular manganese spinels and ferrous sulfide and manganese sulfide, whereas in steels remelted under a flux based on REM oxides and fluorides, inclusions consist mainly of spherical particles of REM silicates and oxysulfides. Due to the action of the REM-based flux, the size of the inclusions decrease, and they increase in absolute number.

Thus, remelting under a flux based on REM compounds promotes the development of finely dispersed spherical inclusions of rare earth silicates and oxysulfides and also reduces the AH content in the alloy because the flux has a low permeability to water. Reduction of AH reduces crack initiation, increases the homogeneity of the alloy, and improves its anticorrosion and mechanical properties.

The advantages of ES alloy, namely, high corrosion resistance, low anisoopy, increased ductility, resistance to hydrogen embrittlement, and stress corrosion resistance make it possible to widen the range of its applications. ES alloys can be used in the production of parts and construction elements that must be able to function in corrosive environments under high loads; railroad rails; stamping tools; seamless, thick-walled pipes for the chemical and the electric power industry; deep-sea equipment; drill bits and diamond tools; etc.

# References

1  G.V. Karpenko, A.B. Kuslitski, and U.I. Babey, "The influence of the density of steel produced by ESR and vacuum Remelting on Its fatigue strength," report to the Academy of Sciences of the UkSSR, No. 8 (1964)

2  M.I. Vinograd and G.P. Gromova, "Inclusions in carbon and alloy Steels," in *Metallurgy*, Moscow, p. 214 (1972)

3  V.S. Ivanova and V.F. Terentiev, "The Nature of metal fatigue," *Metallurgy*, Moscow, p. 456 (1975)

4  G.V. Karpenko, "The tensile strength of steel in corrosive environment," Mashgiz, Moscow, Kiev, p. 188 (1963)

5  G.V. Karpenko, "Adsorptional initiation of corrosion fatigue of metal," report to the AS UkSSR (1951)

6  G.V. Karpenko, "The influence of hydrogen on steel endurance," report to the AS USSR, Kiev, p. 72 (1960)

7  G.V. Karpenko, "The influence of machining on the steel endurance," Mashgiz, Moscow, Kiev, p. 186 (1959)

8  I.V. Kokotailo, A.B. Kuslitski, and U.A. Stavroitov, "Determination of size factor in fatigue tests of contaminated steels," research laboratory report No. 1, pp. 93–95 (1968)

9  G.V. Karpenko, A.B. Kuslitski, and G.I. Zarutski, "Influence of metallurgical heredity on the endurance of roll bearing steels," Vestnik Mashinostroeniya *11*, pp. 36–37 (1974)

10  A.B. Kuslitski, "Nonmetallic inclusions and steel fatigue," Tehnika, Kiev, p. 128 (1976)

11  B.E. Paton and B.I. Medovar, eds., "Electroslag metal," Naukova dumka, Kiev, p. 680 (1981)

12  B.I. Medovar, A.B. Kuslitski, and L.M. Stupak, "The influence of electroslag refining on the corrosion resistance and endurance of metal," Problems of Specialized Metallurgy *16*, pp. 17–21 (1982)

13  I.I. Vasilenko, R.K. Melehov, and N.A. Langer, "The influence of ESR on corrosion cracking resistance of steel 16GHMA," Physico-chemical Mechanics of Materials *2*, pp. 101–103 (1971)

14  V.I. Kovalenko and I.E. Zamostyanik, "Increase of corrosion fatigue resistance of hot worked low-alloy steel by means of ESR." Korroziya i zashita neftegasovoi prom-ti *4*, pp. 2–4 (1981)

15  V.V. Panasyuk, M.M. Stadnik, and V.A. Silovanyuk, "Stress concentration in solids with fine inclusions," Naukova dumka, Kiev, p. 216 (1986)

16  V.M. Finkel, "Physical principles of falure stopping," Metallurgiya, Moscow, p. 360 (1977)

17  G.I. Belchenko and S.I. Gubenko, "Nonmetallic inclusions and steel Quality," Tehnika, Kiev, p. 168 (1980)

18  A.H. Cottrell, "Dislocation and plastic flow in crystals," translated form English by A.G. Rahshtadt, Metallurgizdat, Moscow, p. 267 (1958)

19  J.A. Schulte, "Electrometallury of steel casting," Metallurgiya Moscow, p. 224 (1970)

20  G.M. Itskovich, "Steel deoxidation and modification of nonmetallic inclusions," Metallurgiya, Moscow, p. 296 (1981)

21  L.T. Berezhnitski, V.V. Panasyuk, and N.G. Stashyuk, "Interaction of hard linear inclusions and cracks in a deformed solid," Naukova dumka, Kiev, p. 288 (1983)

22  I.P. Volchok, "Increase of brittle fracture resistance of cast steel, Endurance of Constructions Functioning in Cryogenic temperatures," pp. 64–68 (1985)

23  G.V. Karpenko, I.E. Zamostyanik, and U.I. Babei, "Measurement of stresses in metals on a microscale by means of cell potential," Physico-chemical Mechanics of Materials *5*, pp. 635–636 (1969)

24  G.V. Karpenko, E.M. Gutman, and I.E. Zamostyanik, "Investigation of the microscopic electrochemical homogeneity of metals," Physico-Chemical Mechanics of Materials *3*, pp. 280–286 (1970)

25  "Microscopic electrochemical inhomogeneities of low-alloy steels containing nonmetallic inclusions," Physico-chemical Mechanics of Materials *1*, pp.3–6 (1970)

26  Ya. M. Kolotyrkin, L.I. Freiman, G.S. Raskin, "Local dissolution around nonmetallic inclusions in stainless steel," report to the AS USSR, No. 1, pp. 156–159 (1975).

27  V.V Panasyuk, L.V. Ratych, I.N. Dmitrah, "On the methodology of investigation of crack resistance of cyclically loaded constructional materials in liquid media," Metal Fatigue, Kiev, pp. 50–51 (1981).

28  M.O. Levitski, "The influence of cyclic heat treatment on stress corrosion of cast electroslag steel," Physico-chemical Mechanics of Materials *4*, pp. 50–52 (1980)

29  S.P. Egorov and A.G. Bogachenko, "Melting of thin-walled castings with intricate shapes in a movable crystallizer using a monofilar circuit," Problems of Specialized Metallurgy *13*, pp. 46–49 (1980) p. 46–49

30  M.O. Levitski and B.I. Kovalenko, "Optimization of heat treatment of cast electroslag metal," Electroslag technology, Kiev, pp. 154–158 (1983)

31  V.I. Kovalenko, V.E. Permitin, and K.B. Katsov, "Mechanical and corrosion properties of steel brand 12HN4MDA remelted under rare earth metal based flux composition," report to AS USSR, No. 1 pp. 129–132 (1986)

# 3 Electroslag Remelting as an Efficient Way To Increase the Quality of Alloy for the Production of Wheels and Bearings

M.I. Gasik, U.S. Proidak, and A.P. Gorobets

Much research has been performed through the joint efforts of Dnepropetrovsk Metallurgy Institute (DMI), the E.O. Paton Electric Welding Institute, the K. Libkhnet tube plant, and the "Dneprospetsstal" plant to analyze the quality of train wheel and bearing alloys after electroslag remelting (ESR).

One of the most pressing problems is to provide high-quality, reliabile wheels for all kinds of rail transport, such as high-speed passenger trains, city subway (metro) trains, and rolling stock for the Baikal-Amur Trans-Siberian line.

It is well known that ESR has been used to produce rail steels that have a similar chemical composition to steels used for train wheels [1]. The results of tests on the mechanical properties of rails show that ESR improves the yield point and tensile strength of the alloy through a more general dispersion of pearlite. There are very few published data on changes in the nature and composition of nonmetallic inclusions in rail steel after ESR; accordingly, we analyzed the effect of ESR on wheel steels, on the degree of desulfurization of the alloy, on the change in nonferrous impurity concentrations, and on the composition of nonmetallic inclusions.

Open-hearth and electric furnace steels were melted by ESR. Consumable electrodes of wheel steels melted in the electric-arc furnace were produced with a continuous casting machine (CCM) having a freezer mold with a cross section of $370 \times 370$ mm. Because the open-hearth wheel steel was poured into 12-section molds (4 to 15 tons), consumable electrodes for ESR were produced by rolling the castings to the required cross section ($240 \times 240$ mm). ESR of open-hearth steel was performed in plant OKB-905, and that of electric-arc steel in plant OKB-1065.

During remelting of open-hearth and electric-arc wheel steel under ANF-6 flux with argon gas (6.6 to 8.6 $m^3$/ton), manganese and silicon loss was almost nil. When remelting was performed without a protective atmosphere, manganese and silicon loss was 8% and 16%, respectively (Table 3.1).

Intensive desulfurization of alloys occurs during ESR, which decreases the sulfur content by 40% to 66%, compared to that of the consumable electrodes that were used.

Analyses of the macrostructure of the steels before and after ESR showed that the open-hearth steel contained structures that were etched because of in-

Authors' affiliations: Dnepropetrovsk Metallurgy Institute (DMI), USSR.

**Table 3.1.** Changes in chemical composition of wheels fabricated from open-hearth (OH) and electric-arc (EA) melted steels after electroslag remelting (ESR)

| Melting run (No.) | Melting technology | Chemical composition | | | | | | |
|---|---|---|---|---|---|---|---|---|
| | | C | Mn | Si | P | S | Cr | Ni |
| Initial | EA | 0.55 | 0.73 | 0.33 | 0.011 | 0.012 | 0.13 | 0.04 |
| 1 | EA + ESR | 0.57 | 0.68 | 0.28 | 0.011 | 0.004 | 0.11 | 0.04 |
| 2 | EA + ESR | 0.57 | 0.63 | 0.32 | 0.010 | 0.008 | 0.13 | 0.03 |
| 3 | EA + ESR | 0.55 | 0.65 | 0.26 | 0.010 | 0.004 | 0.12 | 0.04 |
| Initial | OH | 0.59 | 0.76 | 0.26 | 0.010 | 0.027 | 0.11 | 0.15 |
| 4 | OH + ESR | 0.56 | 0.73 | 0.25 | 0.010 | 0.015 | 0.11 | 0.10 |
| Initial | OH | 0.57 | 0.75 | 0.37 | 0.013 | 0.021 | 0.13 | 0.16 |
| 5 | OH + ESR | 0.57 | 0.72 | 0.34 | 0.013 | 0.012 | 0.13 | 0.13 |

homogeneity and the presence of large sulfide inclusions, a rough dendritic structure, regions of relatively low chemical inhomogeneity and nonmetallic inclusions, segregation, and remnants of a shrinkage pipe. In contrast, ESR wheel alloys are homogeneous and dense.

A flaking susceptibility test was performed on wheels produced by different melting methods, and it was determined that wheels made of ESR metal do not flake.

The exceptional reduction of nonmetallic inclusions and gas in the ESR process, as well as the reduction in chemical and dendritic inhomogeneity, produces castings with a dense macrostructure and improved mechanical properties [2]. After ESR, wheel steels are stronger (Table 3.2) and considerably more ductile, because of the purity and crystal orientation of the metal, which helps reduce the size of nonmetallic inclusions and effects a more uniform distribution of these inclusions.

It is well known that impact toughness is one of the most important characteristics of these steels; it has to be no less than 0.2 MJ/m² at 20 °C. To determine the temperature dependence of impact resistance of the wheel risk metal, three

**Table 3.2.** Mechanical properties of 950-mm diameter wheels produced directly from OH and EA steel and from ESR steels

| Melting run (No.) | Melting technology | $\sigma_{hts}$ | $\delta(\%)$ | $\psi(\%)$ | NV |
|---|---|---|---|---|---|
| Initial | EA | 927 | 11.3 | 22.6 | 261 |
| 1 | EA + ESR | 927 | 11.61 | 23.4 | 262 |
| 2 | EA + ESR | 1,010 | 17.4 | 38.0 | 277 |
| 3 | EA + ESR | 990 | 17.2 | 41.0 | 270 |
| Initial | OH | 955 | 12.0 | 20.0 | 270 |
| 4 | OH + ESR | 1,000 | 17.5 | 39.0 | 273 |
| 5 | OH + ESR | 990 | 17.0 | 40.0 | 268 |

Note: According to the requirements of GOST 10791–81, $\sigma_{hts}$ has to be 911 to 1,107 MPa, $\delta$ is no less than 8%, $\psi$ is no less than 14%, and NV = 225 MPa.

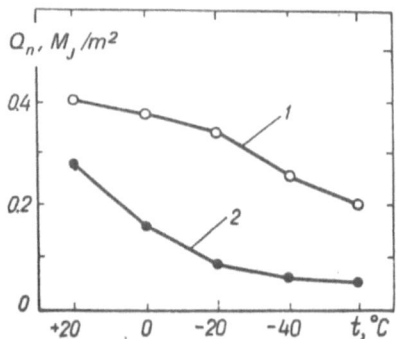

**Fig. 3.1.** Steel fabricated from ESR steel (1) and the starting open-hearth steel (2) in a temperature range of 20 to 60 °C.

samples were tested in a pendulum-type machine. Samples $10 \times 10 \times 55$ mm in size with a type 1 notch, according to GOST 9454-60, were cut radially at the point where the hub merges into the wheel.

Test results show (Fig 3.1) that the impact resistance $(Q_n)$ of ESR steels measured both below and above 0 °C is twice as high as that of ordinary steels. The very low impact resistance of open-hearth steels at −60 °C is caused by increased concentrations of gas, nonferrous metal impurities, sulfur, and non-metallic inclusions.

The data obtained by a thorough analysis of railroad wheel steel refining by remelting consumable electrodes of open-hearth and electric-arc steels were applied to make significant improvements in steel quality by desulfurization, and by decreasing nonferrous metal impurities and nonmetallic inclusions.

In the production of bearings, the most important characteristic of the alloy metal as to quality and service characteristics is its purity, especially regarding the presence of inclusions. It was discovered [1] that, during regular electric-arc melting, the composition of inclusions is determined by the absolute content and relative concentrations of the following dissolved reducing elements: calcium, magnesium, silicon, and aluminum. Inclusion formation can be described by solid-phase mineral formation reactions that occur according to the following relations:

$$m[\text{Ca}] + n[\text{Mg}] + p[\text{Al}] + r[\text{Si}] + q[\text{O}] >$$
$$(m'\text{CaO} \times n'\text{MgO} \times p'\text{Al}_2\text{O}_3 \times r'\text{SiO}_2)\text{l} >$$
$$(\text{Mspn} + \text{Mcor} + \text{Mmul} + \ldots + \text{Mi})\text{s},$$

where Mi is a mineral, that is, an aggregate of oxides that form spinels, corundum, mullite, and other phases in the quaternary phase-diagram.

The control of the absolute content and, more importantly, the activity of the phases makes it possible to regulate the composition of nonmetallic inclusions, as demonstrated by complex tests of commercially produced alloys. The use of electroslag technology (EST) greatly increases steel quality through the control of morphologic changes of inclusions and through a reduction in their number [2].

The successful development of EST in the last decade has changed the economics of bearing steel production greatly. Various technological indexes and quality levels of bearing steels produced by electric-arc melting are now standardized. One of most efficient ways to improve steel quality is to decrease the number of impurities and nonmetallic inclusions [3]. Also, notable results have been achieved from determining the physical characteristics of steels under cyclic loading.

The appearance of a new type of raw material in the steelmaking industry—metallized pellets of direct-reduced iron, with few impurities and a low nonferrous metal content—has permitted the replacement of conventional steel melting in electric-arc furnaces, using metal scrap, by a more advanced technology using these metallized pellets, and also by ESR of the pellets [4].

Melting, using direct-reduced iron (up to 50% of the charge) as a raw material, requires numerous tests to determine the quality of the starting metal produced by open-hearth electric-arc melting and the final product after ESR. The chemical composition of metallized pellets, supplied by the Oskolski Metallurgical plant, met the requirements of TU14-1-3736-8: 90.5% Fe (total); 82.5% Fe (metallic); 1.80% C; 0.007% P; 0.005% S; 4.05% $SiO_2$. Melting was done in 70-ton, electric-arc furnaces equipped with a basic lining and transformers with a specific power rating of 350 kW-A/ton.

When the physical and thermal characteristics of the metallized pellets were taken into account, it was decided to batch load the pellets directly into the hearth. Reduction by aluminum and tapping into ingots weighing 3.3 tons were performed using standard mill techniques. Two ingots from each heating were rolled to produce consumable electrodes $240 \times 240$ mm in cross section; further processing consisted of cleaning surfaces by sandblasting. The ESR of consumable electrodes was run on OKB-1065 machines in freezer molds with a cross section of $415 \times 415$ mm, under ANF-6 and ANF-32 fluxes ([S] = 0.006% to 0.010%). Castings produced by open-hearth electric-arc melting and ESR were rolled into bars with a diameter of 120 to 180 mm. A summary of the results of tests on ESR and open-hearth electric-arc metals are presented in Table 3.3.

Sorting of nonmetallic inclusions into size groups and qualitative metallography analysis were performed according to method III of GOST 1178-70 at a magnification of 280; the results are presented in Table 3.4.

Investigations of the physical properties of metal at different stages of metal processing are of special interest. Consequently, the plasticity of test samples was determined by means of a hot torsion test at the temperature of maximum plasticity, that is, 1,453 °K for steel ShH-15. Test samples were cut from 125-mm diameter round bars from axial and peripheral regions. The results are presented in Table 3.5.

The results of this study show that the nonmetallic impurity content of ordinary alloys is the same as that of an alloy produced by melting 50% metallized pellets in the charge. In addition, the most notable improvements in steel char-

**Table 3.3.** Inclusions in EA steel ShH-15 before and after ESR

| Melting Technology | Fraction of metallized pellets in charge | Melt (No.) | Rating according to GOST 801–78 (points) | | | | | |
|---|---|---|---|---|---|---|---|---|
| | | | Outgoing inspection | | | Laboratory inspection | | |
| | | | O | S | SP | O | S | SP |
| OEA | 0 | 3 | 2.26 | 2.30 | 2.11 | 2.05 | 2.21 | 2.47 |
| | 0.20 | 11 | 2.0–2.5 / 2.19 | 2.0–2.5 / 2.24 | 2.0–2.5 / 2.03 | 1.5–2.0 / 1.8 | 1.0–2.0 / 1.80 | 1.5–2.5 / 1.87 |
| | 0.30 | 23 | 2.0–2.5 / 2.1 | 2.0–2.5 / 2.1 | 1.5–2.0 / 1.90 | 1.0–2.0 / 1.5 | 1.5–2.5 / 1.87 | 1.0 / 1.0 |
| | 0.40 | 8 | 2.0–2.5 / 2.25 | 2.0–2.5 / 2.25 | 2.0–2.5 / 2.25 | 1.0–2.0 / 1.5 | 2.5–3.5 / 2.9 | 1.0–2.0 / 1.55 |
| ESR | 0.30 | 6 | 0.5–1.0 / 0.83 | 0.5–1.0 / 0.83 | 1.0–1.5 / 1.16 | 0.5–1.0 / 0.75 | 0.5–1.0 / 0.75 | 0.5–1.0 / 0.75 |

Note: Above the line, minimum and maximum; below the line, average; O, oxides, S, sulfides, and SP, spheroids.

**Table 3.4.** Nonmetallic inclusion content and distribution according to size in steel ShH-15 melted with a charge of 40% metallized pellets (above the line) and with futher ESR (below the line)

| | | | Weight of inclusions/100 mm³ of cross section (%) | | | | |
|---|---|---|---|---|---|---|---|
| | | | Distribution by size ($\mu$m) | | | | |
| Nonmetallic inclusion type | Volume (%) | Total | Less than 1.0 | 1.01–2.5 | 2.51–5.0 | 5.01–10 | More than 10 |
| Angular | 0.0041 | 597 | 10.2 | 52.4 | 27.2 | 10.2 | 0 |
| Oxides | 0.0031 | 566 | 9.2 | 50.0 | 39.2 | 1.8 | 0 |
| Nitrides | 0.0014 | 172 | 0 | 41.3 | 52.9 | 5.8 | 0 |
| | 0.0002 | 30 | 0 | 33.3 | 66.7 | 0 | 0 |
| Sulfides | 0.0209 | 1384 | 1.4 | 44.5 | 35.0 | 14.6 | 4.5 |
| | 0.0144 | 565 | 1.8 | 42.8 | 33.2 | 21.4 | 30.8 |
| Spheroids | 0.0088 | 919 | 3.3 | 46.1 | 38.5 | 11.0 | 1.1 |
| | 0.0018 | 323 | 9.3 | 59.4 | 25.1 | 6.2 | 0 |
| Oxysulfides | 0.0025 | 91 | 0 | 0 | 33.0 | 67.0 | 0 |
| | 0.0055 | 172 | 5.0 | 11.6 | 29.7 | 52.9 | 0.8 |
| Total | 0.0377 | 3163 | 3.5 | 45.0 | 35.5 | 13.8 | 2.2 |
| | 0.0250 | 1656 | 4.5 | 45.1 | 34.5 | 14.6 | 0.3 |

**Table 3.5.** Ductility of steel ShH-15 melted with additions of metallized pellets to the charge, and after ESR

| | Turns before fracture (No.) | | | |
|---|---|---|---|---|
| Melting technology | Rod | Center | Edge | Anisotropy coefficient |
| Ordinary charge | A | 39 | 57 | 1.46 |
| | S | 38 | 53 | 1.37 |
| | N | 33 | 50 | 1.52 |
| Charge with 30% pellets | A | 48 | 63 | 1.31 |
| | S | 46 | 60 | 1.30 |
| | N | 44 | 57 | 1.29 |
| Charge with 30% pellets + ESR | A | 55 | 62 | 1.12 |
| | S | 53 | 57 | 1.07 |
| | N | 51 | 61 | 1.20 |

acteristics are provided by the subsequent ESR of the alloy. Integrated steel-making, as described here, not only yields steels of a lower than average impurity, it also controls the composition of the inclusions formed. Moreover, it optimizes grain orientation, and improves the properties (tensile strength, ductility, impact toughness, isotropy) of the steel.

Thus, ESR increases the capabilities of existing steelmaking technologies and also makes possible new methods of bearing and wheel metal production.

# References

1  I.G. Uzlov, M.I. Gasik, and A.T. Esaulov, "Wheel steel," Tehnika, Kiev, p. 168 (1985)
2  M.I. Gasik, J.A. Schulte, and A.P. Gorobets, "Physico-chemical patterns of spheroidization of inclusions in bearing steel," Izv. Vus. on Ferrous Metallurgy 5, pp. 10–15 (1983)
3  B.I. Medovar, M.I. Gasik, and I.G. Uzlov, "Influence of ESR on wheel steel quality and properties," Problems of Specialized Metallurgy 4, pp. 27–30 (1986)
4  B.I. Medovar and L.M. Stupak, "ESR of metallized pellets, Ferrous metals industry information," Steel Melting Industry 5, p. 26 (1984)

# 4 Electroslag Remelting in the Development of New Steels for Fabricating Parts and Bearings

V.S. Levitin, V.S. Kropachev, and E.G. Zaharov

Since its founding in 1965, the Research Laboratory of the Zagorsk branch of VNIIPP has been developing and improving bearing and tool steels. The laboratory is involved in the pilot production of batches of alloys to determine their physical properties and other properties specific to their proposed service as bearings. It has also undertaken the production and testing of pilot batches of alloys for special-purpose bearings.

The laboratory is equipped with radiofrequency-induction (RFI) furnaces with crucible capacities of 160 and 450 kg to melt metal and to cast consumable electrodes into cast iron-lined split molds, vacuum induction furnaces with a crucible capacity of 50 kg to produce specialized alloys, furnace installations to remelt electroslag (ES), and heat treatment furnaces and instruments to analyze alloys.

Cast consumable electrodes 0.07 to 0.3 m in diameter and 1.5 to 2.1 m in length are remelted without preconditioning in a hermetic ES furnace, type U-102, built in the E.O. Paton Electric Welding Institute, or in single-phase open-hearth furnaces, which is based on A-550U ESR furnaces. The remelted castings are 7 to 30 cm in diameter, and weigh between 20 and 300 kg. The open-hearth furnaces use either a solid or a liquid charge; in the latter case, a graphite ladle-crucible and a permanent electrode are used. There are provisions for electroslag remelting (ESR) in an inert gas atmosphere.

A wide range of pilot steel batches have been produced by ESR. Table 4.1 lists the carbon- and alloy-bearing steels that have been analyzed after ESR. Examples of carbon and alloy steels that were developed in ZFVNIIPP and VNIIPP are now widely used in industry; they are as follows: ShH15SM-Sh (DI36-Sh), 40H11M3F-Sh (EP890-Sh), OH14N8M2S3-Sh (EP821-Sh), 8H4M4V2FT-Sh(DI43-Sh), and an alloy 38HNVYu-VI. New technologies have been developed to produce and analyze ES tool steels for hot and cold working; high-speed steels; oxidation-resistant, chromium–nickel–silicon steels for heat-treating equipment; 36NHTYUM8 and 40KHNM steels; and specialized cast irons for high-precision laps. Over 250 steel types are produced by ESR.

During the manufacture of pilot batches of the carbon and alloy steels mentioned above, a special effort was made to follow the ESR technology used in metallurgical plants. The further study of rolled rods permits the determination of alloy characteristics, which are then fed back to the production line. By fabricating pilot batches under ESR conditions close to those used in industry, the

Authors' affiliations: VNIIPP Department in Zagorsk, USSR.

**Table 4.1.** Electroslag remelting of carbon- and alloy-bearing steels

| Carbon and alloy designations | Melted composition number | Steels used in production |
|---|---|---|
| ShH15Yu2, ShH15SYu | 18 | ShH15SM-Sh (DI36-Sh) |
| ShH15M, ShH15B | 6 | |
| 11H18MS, 11H15M4F, 11H16M3V3, 14H15M4F | 23 | 130H15M4F3 (3I 120-BD) |
| 50H18M, 60H18M, 95H18 | 11 | 50H18AM-ShP (EP981-ShP) |
| 6H4B9F, 7H5V8MF, 7H5M3V8F | 16 | 8H4V9F2-Sh (EI347-Sh) |
| 55H5M3V2FT, 6H5M3V2FYu | 11 | 8H4MV2FT-Sh (D143-Sh) |
| 00H12N7K15, 00H13N8K10M3 | 14 | 00H14N8M2S3-Sh (EP 821-Sh) |
| 50H12S2F, 40H13ST, 40H10S2M | 15 | 40H11M3F-Sh (EP 890-Sh) |
| 20HGNF, 20HZGF, 18HN2MYu, 20H2GNST | 18 | 20HZSGMF-Sh |
| 40HNSYu, 36HNV2M610 | 24 | 38HNVYu-VI (EP940-VI) |
| 30H35MV, 27H35M3 | 7 | 23H32M2-Sh |

period between the development of a new material and its use in industry can be greatly reduced. This approach short-circuits the fabrication cycle of modification and testing that usually characterizes pilot production of alloys by open-hearth melting technology.

The specialized techniques used to produce bearings often require solving unusual problems not encountered in conventional ESR. One such problem is the fabrication of small batches (up to 50 kg) of deformation-resistant alloys (23H32M2-Sh) for the manufacture or 1-in. diameter roller and ball bearings for rolling contact bearings. These alloys are known for their hardness after annealing (35 to 45 $H_{RC}$), their low impact resistance, and their low machinability. That is why the fabrication of small rollers and balls from large-diameter ES castings (0.1 to 0.3 m), which contain large primary carbides and have a high carbide inhomogeneity even when they are remelted at the lowest possible solidification rates, is both complex and difficult. The use of ESR to produce thinner rods—with lengths of 0.7 to 1.3 m and a diameter of 0.03 m—from 0.1-m diameter electrodes solves two problems: obtaining a finely dispersed carbide phase by intensive cooling and producing an appropriately shaped product for fabricating rollers and balls. The installation for the production of small-diameter rods uses a liquid charge and its drawing rate is regulated automatically. Molten metal and flux levels are determined by thermocouples inserted into the parting line of the variable cross section freezer mold. Roller bearings obtained from such preforms have been found to be successful in the field.

The production of bearing races from deformation-resistant alloys, like 23H32M2, was a joint effort undertaken with TsKTBMM of the "Krasnoe Sormovo" factory to demonstrate the possibility of manufacturing hollow preforms by ESR in a magnetic field. Analysis of these preforms and their comparison with solid castings produced by ESR made it possible to determine the dependence of the carbide-phase dispersion on the casting diameter. The quantitative dependence of the average size of large primary carbides in deformed 11H18M-

Sh steel on the casting diameter made possible the regulation of carbide phase size and form in steels to produce improved bearings.

A second example of a nontraditional problem in ESR is the development of technology for the production of sulfur-containing cast iron (0.15 to 0.35% sulfur content) for high-precision lapping tools. An ESR process used to produce special gray cast iron was employed; a 1:1 mixture of conventional fluxes (ANF-6 and AN-26) was used to stabilize the electrical characteristics during remelting, to obtain high-quality castings. Numerous tests performed at ZFVNIIPP, the "Kalibr" plant, the Istrinsk department of VNIIEM, VNIIASh, and other enterprises showed that ES-cast iron in uniquely suited for the fabrication of laps for use on metal parts of various hardnesses because of its high homogeneity; its fine, pearlite structure; its improved ability to hold various size abrasives; and its enhanced wear resistance.

Alloying cast iron with sulfur and manganese greatly improves its properties as a lap material because its spheroidal inclusions of manganese and iron sulfides act as solid lubricants and as location sites where abrasive particles lodge. A new process was suggested for ESR for sulfur-containing cast iron, wherein at least 0.15% sulfur is retained and is uniformly distributed throughout the casting, thus guaranteeing high-dimensional stability and better physical properties. As a result of tests run on different ESR processes (in particular, those in which sulfur-containing iron is used during remelting), it was found that the best properties are obtained by ESR when a mixture of fluxes (ANF-6 and AN-26; 1:1) are used and 30% $SiO_2$ is added. In this method, 60% of the primary sulfur content is retained in the consumable electrode to provide more homogeneity and to improve composition and physical properties. The use of ESR sulfurized cast iron improved the performance and wear resistance of lapping tools by 20% to 30%.

Considerable research was done to determine the quality of ShH15-Sh steel. As long ago as 1967, it was discovered during studies of flux basicity that when acid K-3 fluxes were used, the concentration of silicate and the number of spherical inclusions observed on a polished cross section increased by one order of magnitude, compared to results obtained with the ANF-6 flux. However, highly basic fluxes (ANF-7) and a liquid charge yield a high-density, defect-free macrostructure and a twofold decrease in nonmetallic inclusions. The regions containing spheroidal inclusions (the worst kind), however, increased in size. The results of these studies demonstrated the necessity of using either two ESR steps, carried out under different fluxes, or vacuum-arc melting after ESR, to reduce impurities in the steel. As is well known for steels used in ball and roller bearings, it is of the utmost importance to control the number of inclusions in order to obtain very smooth surfaces and precise dimensions. These parameters determine the bearing starting torque, the level of vibration, and other operating characteristics.

In addition, pilot batches of ShH15-Sh steel were produced by a newly developed technology that uses pure charge materials (direct-reduced iron, electrolytic chromium, crystalline silicon, and metallic manganese) to reduce the content of the so-called residual elements (tin, lead, zinc, antimony, arsenic, bismuth, etc.) that have low melting points and low solid-state solubilities, as well as to eliminate such elements as aluminum and titanium, which may be present as refractory-compounds. A study of alloys produced by this new technology demonstrated significant improvements in operating characteristics.

After the most effective means of scrap recycling to obtain additional sources of raw material was determined, a new technology for remelting ShH-15 scrap was developed. It uses RFI melting in air with a charge of up to 20% machining scrap. Steel melted in this manner is designated ShH15-Sh and conforms to GOST 801-78. It was also possible to increase the scrap fraction to 60% and still obtain steels with inclusion contents that satisfied GOST 801-60 for the ShH-15 designation and had, in addition, increased structural homogeneity, which improves service characteristics.

One example of research findings that were subsequently introduced into factory production was the development of ESR of the high-alloy, heat-resistant steels 8H4M4V2FT-Sh (DI43-Sh) and 9H6M3F3AGST (EK41-Sh). By means of this ESR process, it was possible to limit titanium concentration to a narrow range (0.1% to 0.3%) as well as to retain the nitrogen that had been introduced for austenitizing and additional hardening. In recent years, there has been a tendency to use less expensive molybdenum–tungsten steels that have lower hardening temperatures than plain tungsten steels. The low-solubility, carbo-nitride particles in these steels are particularly efficient in retarding grain growth. At low titanium concentrations and in the absence of aluminum, titanium oxides can be produced by flux reduction. ESR under a high-flouride flux with up to 20% titanium oxides keeps titanium melting losses at a minimum, retains nitrogen, and produces highly homogeneous castings. The addition of titanium to 8H4M4V2FT-Sh steel increases the range of hardening temperatures from 20 to 60 °C and increases the mean bearing lifetime ($L_{90}$) from 190 to 1,000 h.

The ZFVNIIPP research laboratory has shown that using ESR to create specialized new materials can reduce the cost of research, speed up the development cycle, and facilitate the introduction of new, more efficient materials.

# 5 Electroslag Remelting of Steel Slabs for the Fabrication of Thin Strips

Yu.M. Kamenski, B.M. Romanov, O.N. Drushinina,
N.N. Perevalov, and R.V. Kakabidze

In the course of modernizing the "Serp i molot" metallurgical plant, it was decided to start the manufacture of medium sized ingots and slabs by electroslag remelting (ESR) in order to supply new billet mills with high-quality steel melted directly at the plant. The modernization was aimed at refitting the plant with new rolling mills and also initiating production of small-sized billets and hot-rolled strip and sheet.

The E.O. Paton Electric Welding Institute proposed a remelting technology that would utilize a two- or four-strand, movable, freezer mold with four consumable electrodes that are melted in a common slag bath and powered by a bifilar circuit. For the practical realization of this technology, it was decided to update the existing R-951-type installation that was used for remelting. In addition, it was necessary to build a new flux melting installation, since it is impossible, in principle, to develop a process based on a solid flux in multielectrode ESR equipment, as is sometimes done with a stationary freezer mold and a single-phase power supply. The main advantage of using a liquid charge in multistrand equipment is that it is no longer essential to provide water cooling for the bottom plate; this increases the level of safety during the initial stages of the process.

To produce high-quality steel ingots and slabs by means of ESR that uses a movable freezer mold, it was necessary to develop an automatic control system (ACS) to regulate the rate of withdrawal of the freezer mold, which should closely follow the solidification rate of the ingots. The ACS is programmed to control the level of the metal–slag boundary, which is measured by thermal sensors that detect sharp changes of heat flow from the molten metal and the slag baths. This type of ACS was introduced at the ESR shop of the "Serp i molot" metallurgical plant in Moscow, which had the necessary operating speed, reliability, and selectivity.

The thermal level sensor consists of a cylindrical, water-cooled copper capsule contianing two miniature arrays of thermocouples connected differentially, with the hot junctions inserted into the capsule bottom. The two thermocouple arrays are placed one above the other in the direction of motion of the freezer mold.

The introduction of these multistrand ESR furnaces with movable freezer molds for commercial manufacturing allows the remelting of wide assortment of steels, including complex alloy steels, and particularly corrosion-resistant austenitic steels that do not contain titanium. In collaboration with the E.O. Paton

---

Authors' affiliation: "Serp i molot" Metallurgical Plant, Moscow, USSR.

Institute, new fluxes were developed to change the electrical resistivity of the slag layer to correspond to the liquid temperature of the melt. High-quality ingots were obtained by using combinations of slags.

When ERS is performed in a movable freezer mold, there arise problems due to variations in the thickness of rolled and, especially, forged electrodes. Such variations lead to differences in the weight of ingots even when the current is evenly divided among the strands and, in practice, necessitate halting the melting process because each strand is of a different length.

A more advanced method of fabricating consumable electrodes is now in use. The electrode alloy is melted in arc furnaces and subsequently cast in horizontal, low-pressure machines (HLPM), including experimental units.

Existing ESR plants were studied to determine if the mathematical models of the process developed by the VNIIETO could be improved and to use these models to calculate wall temperature profiles of freezer molds with square and rectangular cross-sections [1].

Using these calculations, the VNIIETO designed and built movable multi-strand freezer molds. These freezer molds successfully passed industrial tests in which low- and high-alloy steel electrodes were used. The test results showed that the operating current $I_o$ depends on the electrode diameter $d$, which is given by the relation $I_o = 300d$, this differs form that observed earlier for ESR in which a single consumable electrode and a stationary freezer mold were used [2].

In commercial tests, which demonstrated the possibility of melting square and rectangular cross-sectional ingots, electroslag slabs were cast from 08H18Sh10 steel with cross sections of $150 \times 400$ mm and lengths of 2,000 mm. These slabs were subsequently formed into hot-rolled coils $350 \times 400$ mm in cross section for further use in strip production. The ESR steel was highly ductile. After heat treatment and etching, the surface of the rolled sheet was rough, with remnants of unetched scale on both sides, but no seams were observed. Analysis of the chemical composition showed a very slight difference in content between the head and tail ends of the coil and in the content of key elements in samples taken from two melting runs. The chemical composition (%) of the coil after ESR was determined. First melting run: C, 0.05; Mn, 1.13; Si, 0.62; S, 0.004; P, 0.016; Ni, 10.10; Cr, 17.0; Cu, 0.21; Al, 0.60; and second melting run: C, 0.03; Mn, 1.16; Si, 0.56; S,0.004; P, 0.018; Ni, 10.90; Cr, 17.0; Cu, 0.21; Al, 0.25.

A strip 0.25 mm thick (with a tolerance of $-0.02$ mm) and 400 mm wide was fabricated from the hot-rolled sheet in a cold strip mill used to produce stainless steel strip.

Trace amounts of tungsten, molybdenum, titanium, and vanadium were detected in steel from these melting runs. Concentrations of oxygen and nitrogen were 0.011% to 0.021% and 0.049% to 0.066%, respectively.

The ESR steel exhibited high ductility during conventional cold rolling. Mechanical tests demonstrated that peened strip, 0.25 mm thick had a yield point of 935 to 1040 MPa, a tensile strength of 1,060 to 1,080 MPa, and an elongation of 5.0% to 6.4%.

A comparsion of the mechanical properties of samples across and along the rolling direction taken after bright annealing ($T_{max} = 1,070$ °C; speed, 18 m/min) revealed only a slight anisotropy in ESR steel. Longitudinal samples had a yield stress of 250 to 255 MPa, a tensile strength of 610 to 630 MPa, and an elongation of 53.0% to 56.0%, whereas the corresponding values for transverse samples were 240 to 250 MPa, 590 to 600 MPa, and 53.0% to 57.0%.

A structural microanalysis of strip samples revealed highly homogeneous austenite grains, 7 to 8 on the size scale, and small nitride particles, 0.5 on the size scale. No precipitates within the grains or on the grain boundary were detected. The surface roughness was no greater than 0.09 $\mu$m.

The high quality of cold-rolled strip fabricated from ESR slabs produced in movable, tapered, freezer molds is evidence of the successful modernization of steel production in "Serp i molot" factory, which achieved it goal of production sheet and strip from steel melted in the plant.

# References

1  G. Omura, M. Bakabayasy, and T. Hosada, "The analysis of heat transfer during ESR," Naukova dumka, Kiev, pp. 180–202 (1973)
2  B.E. Paton and B.I. Medovar (eds.), "Electroslag furnaces" Naukova dumka, Kiev, p. 415 (1976)

# 6 Computer Automatization of Electroslag Remelting Furnaces

V.I. Mahnenko, E.D. Gladkii, Yu.A. Skosnyagin, and V.I. Zayats

The development of automatic control systems (ACS) for commercial electro-slag remelting (ESR) has been conducted at the E.O. Paton Electric Welding Institute since 1978. During this time, considerable experience has been gained in constructing this kind of system, along with analyzing how such systems function in industrial use. In this chapter three examples of ACS are given to illustrate the various control structures that were developed for specifiic ESR furnaces and technologies. These ACS were built in the Kramatorsk Manufacturing Association (MA) NKMZ, in the Kiev MA "Bolshevik," and in the Zhdanov metallurgical plant "Azovastal."

The goal in creating the above systems was to guarantee steel quality. The control techniques that were applied depended on the details of the remelting process in each case; the primary goal, however, was to create conditions for the optimal refining and subsequent solidification of the steel. Because it was not possible to make measurements that would allow direct control of processes, efforts were directed toward developing automatic regulation systems (ARS) so that it would be possible to obtain a general idea of casting quality by determining several control parameters. Tests on different versions of ESR control systems revealed that the most appropriate parameters for this purpose are the ingot solidification rate and the slag layer electrical resistance. At present, all commerical ESR furnaces in the Soviet Union are equipped with ARShM-T controllers that regulate the solidification rate and the bath resistance by computer [1, 2]. However, the use of this type of regulator still did not solve many of the problems involved in obtaining the necessary high quality of ESR steel. This inadequacy arises beacuse the ESR process is characterized by a high thermal inertia, and thus, furnace control systems are required that are not only capable of reacting to deviations of the control parameters from their setpoints, but also are capable of predicting effects of such deviations on the casting quality and of altering the setpoints through local ARS, if necessary. An important point in the design of an ESR control system is the provision for interactions between various control loops. First, for a given type of steel and casting shape, it is essential to choose appropriate electrical parameters for remelting (slag layer current and voltage, etc.) as well as the mechanism for cooling the freezer mold and the bottom plate. In addition, maintaining an optimal chemical composition of the molten metal pool and layer baths throughout the melting run is of great impor-

Authors' affiliation: E.O. Paton Electric Welding Institute, Academy of Sciences, UkSSR.

tance, since it eliminates detrimental impurities (sulfur, dissolved gases, non-metallic inclusions). In order to improve ingot quality in the course of ESR, various reducing agents are introduced into the melt by means of an injector, and the slag mixture or some of its constituents are added to maintain the proper volume and chemical composition of the slag layer.

In order to melt both solid and hollow ingots in a movable freezer mold, it is necessary to regulate the rate of advance of the furnace mechanism as well as the solidification rate of the ingot. To address all these problems of ESR furnace regulation, it is necessary to construct systems that allow one to be able to choose and maintain optimal melting conditions. The most important task is the creation of an ACS for high-capacity furnaces, in which new techniques are tested by repeated melting runs, which are very expensive and labor intensive. For this reason, ESR furnace control systems based on minicomputers and microprocessors are a very promising, useful development.

In the creation of ACS at the E.O. Paton Institute, considerable attention was paid to various means of increasing the efficiency of ingot production, such as improving steel quality and reducing the cost of further manufacturing operations; taking off less of the crop ends from the top and bottom of the ingot; reducing the cost of developing processes for new types of steel or castings of a different size and shape; increasing productivity while maintaining steel quality; and reducing power consumption. In addition, the control system had to guarantee consistency in ingot quality by eliminating the effect of operator subjectivity, as well as freeing the operator from the routine tasks of controlling melting conditions and documenting the process.

We now give a more detailed description of each of the above three commercial systems.

## The Automatic Control System of ESR in the Manufacturing Association "Novokramatorski Mashinostroitelnyi Zavod"

The ESR shop of the MA NKMZ is a specialized department for the production of forgings of structural, roller, and other steels. The shop processes can be divided into three main groups: auxiliary operations, remelting of steel into ingots, and finishing operations. The ACS that was built for this factory was designed for centralised control of the entire shop. This chapter describes the part of the system that regulates 6ESR-20SV and ESR-10 furnaces. Since the main mill products are large ingots, produced in small batches, for industry, the ACS has a dual purpose: to help the process engineer develop new melting schedules, and to control mass porduction of ingots according to the technology chosen. Another requirement for this control system is to maintain preset melting conditions during a long process run (tens of hours). These requirements were taken into account when the system was constructed.

The ACS was designed to be fail-safe; that is, it would ensure that the shop equipment continued to operate reliably in case of computer system failure through the deployment of local automatic control devices. The system is hierarchical, with two command levels: the upper level is controlled by an SM-2 computer, whereas the lower level consists of several local controllers. The system orgainzes the collecting and processing of data from the ESR furnaces; these data have about 200 parameters, each of which is checked on critical values. This information is filed with the process documentation. The system analyzes the collected data, determines the optimal melting conditions, and provides for the necessary regulating operations.

The analysis of the optimal melting regime is based on mathematical modeling of the physicochemical processes occurring during melting; these processes usually cannot be monitored directly. This approach makes it possible to determine the process conditions at a given time, as well as to predict furture condi-

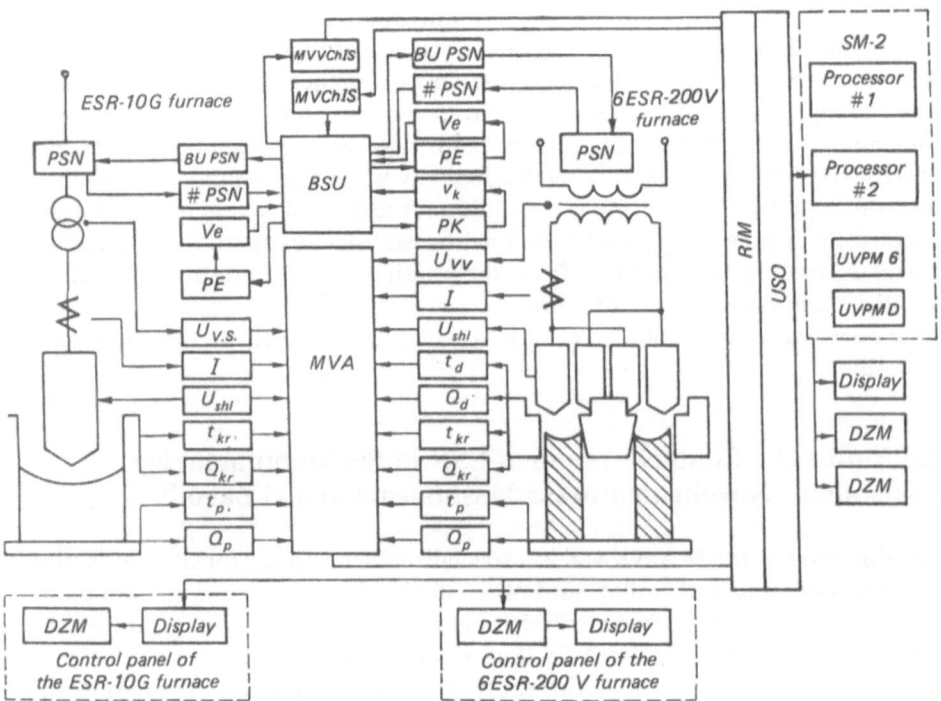

**Fig. 6.1.** Block diagram of the ESR automatic control system in the NKMZ Manufacturing Association. Key: MVVChIS, digital input port; MVChIS, digital output port; BSU, contactless controllers; MVA, analog input port; PSN, furnace voltage switch; BU PSN, furnace voltage controller; # PSN, the number of the voltage controller; PE, electrode driver; PK, freezer mold driver; DZM, printer; RIM, interface multiplexer; USO, computer interface; UVPML, magnetic tape storage; UVPMO, magnetic disk storage; $v_e$, electrode travel rate; $v_k$, freezer mold travel rate; $U_{v.s.}$, supply voltage; I, melting current; $U_{shl}$, slag layer voltage; $Q_d$, water consumption for mandrel cooling; $Q_{kr}$, water consumption for freezer mold cooling; $Q_p$, water consumption for bottom plate cooling; $t_d$, temperature of outlet water from the mandrel; $t_{kr}$, temperature of outlet water from the freezer mold; and $t_p$, temperature of outlet water from the bottom plate.

tions. For this purpose, complex calculations that estimate the parameters of ingots remelted under the intended control conditions are performed. If necessary, these parameters are corrected to optimize the melting conditions.

The control system has, in addition, auxiliary functions that are used to run preliminary trials of new technologies and to determine appropriate melting regimes without undertaking actual melting runs. In the mathematical modeling of a proposed new process, both software and hardware are used. Such modeling of the control system response using a new process technology makes it possible for the engineer to estimate the likely range of melting parameters before undertaking any melting runs. A block diagram of the control system is given in Fig. 6.1.

The system is based on an SM-2 mainframe computer, which is connected with peripheral equipment and input–output devices at each furnace. It accepts inputs from various process sensors as well as terminals and control panels. Control commands are either displayed directly on local regulating systems or on terminals attached to the furnace.

## The ESR Control System in the MA "Bolshevik"

The ESR mill of the MA "Bolshevik" is equipped with YSh-116 and ESR-2.5 installations. The mill produces small batches of hollow ingots of various shapes and sizes and complex castings for use in the chemical industry. The size and shape of the ingots are such that it is necessary to have computer control over the electrical power to the furnace; the slag layer resistance; the ingot solidification rate; and the addition of flux, reducing agents, and alloying compounds. Each parameter is kept within the required range by local controllers. A computer provides the most efficient coordinated control of the interdependent parameters, which are determined by complex, time-dependent, nonlinear equations..

When formulating the task for any of the local regulators, it is necessary to analyze causes of deviations from setpoints and to correct other control loops, if necessary.

Systematic variations and random variations are the most important perturbations that can occur during melting. Examples of systematic variations are changes of the electrical characteristics of the furnace circuit, caused by a reduction of the electrode length, and a change in their location; changes in the volume and chemcial composition of the slag layer; and changes in the thermal characteristics of the electrode–slag–ingot system. Examples of random variations are random transients on the power line and variations in the chemical composition and the dimension of the electrodes. The overall control system, which consists of multiple feedback loops, was designed to minimize the effects of both types of perturbations.

As was done in the MA NKMZ, a fail–safe type of control system was adopted at the MA "Bolshevik." Control is also hiararchical, with two command levels: the upper level is provided by an SM-2 computer complex; the lower level, by a group of local regulators. The system organization resembles that of the ACS built in the MA NKMZ (Fig. 6.1). We discuss system functions and the interconnections of the various components in more detail below.

Data-processing functions of the system consist of measuring and checking deivations from the setpoint of 10 critical parameters of each furnace and also calculating and checking 7 more parameters. All 17 parameters are entered into memory and can be printed out if desired. Seven of them are displayed on monitors. In addition, the data obtained by checking for deviation from the set values can warn the operator of possible hazardous conditions during the melting process.

In small-scale batch production, it is frequently necessary to input additional data, and thus, great attention was paid to the interaction between the operator and the computer. The system contains several levels of dialogue: with the steelmaker, the engineer, and the operator. In each case, the dialogue helps to solve different problems depending on the information available in the memory. The dialogue with the steelmaker is the lowest level and takes place by way of a portable terminal. Before the melting run, the steelmaker enters data concerning the type of melting run, parameters, and the personnel who will perform the run. After these data have been entered, the computer checks furnace readiness and then displays the recommended initial settings for the melting run. If the furnace conditions are appropriate, the computer gives permission for the flux to be loaded. From this point to the end of the melting run, the entire pocess is controlled automatically without operator intervention.

The dialogue with the engineer can occur during the preparation period and the melting run itself. Through this dialogue, it is possible to compare the parameters being measured with those stored in the database; for example, the idling voltage, the solidification rate of the ingot, and the frequency of updating of information. Accordingly, this type of dialogue is convenient for entering data for new melting schedules.

Through dialogue with the operator, the system provides for the organization of data storage and the determination of the servicing order of ESR furnaces.

Overall, the control system performs the following functions: stabilization of the slag layer resistance; process optimization by maintaining a constant solidification rate while minimizing power consumption; and control of the solidification rate, the melting power, and the rate of injection of additives, flux, reducing agents, and alloying compounds.

The system structure is based on commercially available ASVT-M and GSP devices. In addition, there are separate interfaces with the ARShM-T regulator; these were developed in the design and construction department (OKTB) of the E.O. Paton Institute. The hardware consists of various process sensors with their associated data acquistition converters, local control systems for ESR equipment, normalizing converters, control panels, and an SM-2–based computer complex (Fig. 6.2).

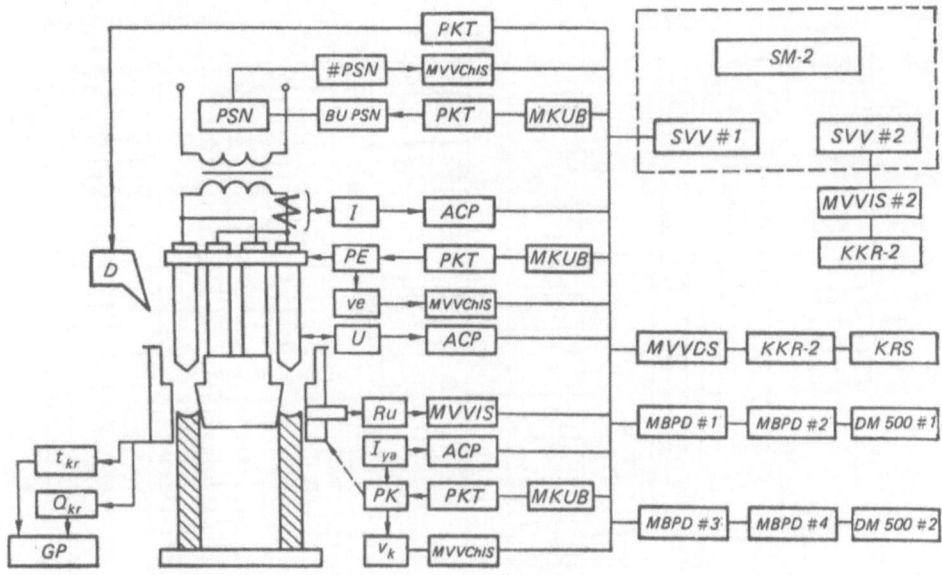

**Fig. 6.2.** Block diagram of an ESR automatic control system at the "Bolshevik" Manufacturing Association. Key: RU, level regulator; GP, joint converters; PKT, digital/analog converter with current output; ACP, analog/digital converter; MVVIS, initial signals imput–output unit; MKUB, contactless digital controller; SVV, input–output module; KKR, reverse comand switch; MVVDS, digital input–output; KRS, relay communication switch; MBPD, data transmitter; DM500, terminal; and D, additive injector.

## The ACS of the "Azovstal" Metallurgical Plant

The electroslag (ES) shop of the "Azovstal" steel mill is directed toward producing thick steel plate from structrual steels, high-strength steels, and other steels. The shop is equipped with ESR-20BG-I2 furnaces designed to produce ingots of various shapes and sizes. In order to obtained high-quality steel with this type of equipment and melting process, it is necessary to combine both of the above approaches to ESR process control; that is, the control system should provide for the coordinated programmed control of process parameters together with regulation of the ingot melting condition.

The system has a two-level, distributed control organization [3]. A block diagram of its structure is given in Fig. 6.3. At the lower control level, each furnace is equipped with a local control system based on the SM-1803 microprocessor, which is mounted directly in the furnace control panel. This lower-level control system performs the following tasks: the acquiring and the preliminary processing of data, including measuring both analog and coded digital signals, along with determining the calculated values of the process parameters; interacting with the production personnel, inputting initial data for melting by the technician, displaying information on the process parameters, and recommending pro-

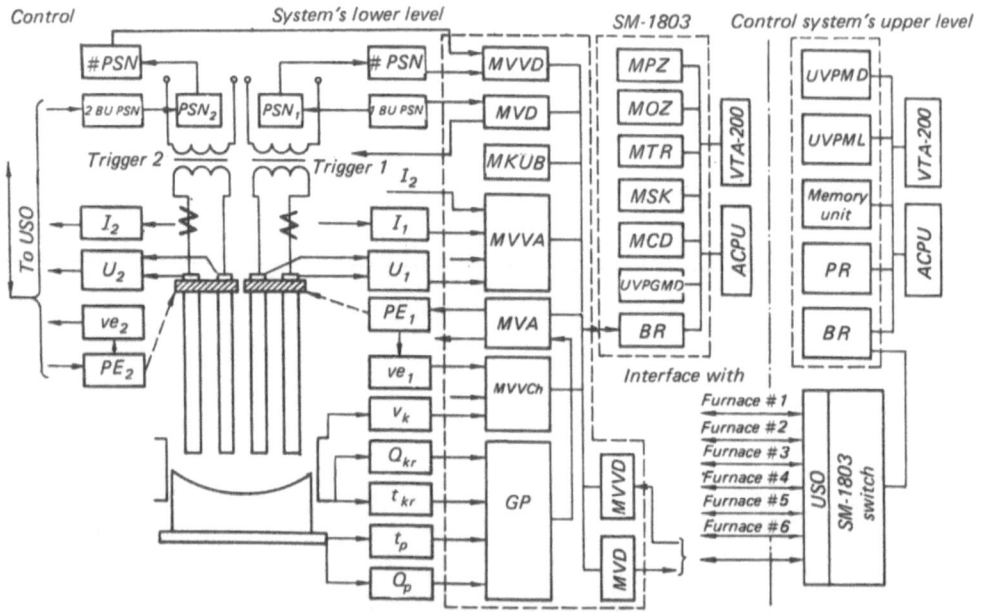

**Fig. 6.3.** Block diagram of the ACS of the 20BG-12 ESR furnaces at the "Azovstal" metallurgical plant. Key: MVVD, digital signal input module; MVD, digital signal output module; MVVA, analog signal output module; MVVCh, digital data input module; GP, joint converters; VTA, display terminal; ACPU, alphanumeric printer; USO, computer interface; and UVPGMD, floppy disk storage.

cess regulation during a melting run; processing control governed by the main parameter values; organizing melting schedules, including determining the initiation of the melting of the charge and transiting to steadly-state melting and feeding; filing and printing out the necessary documentation and melting report; interacting with the upper control level, including creating and transmitting data arrays containing the melting parameters and receiving and processing data from the higher-level system.

The control functions of the lower-level system are disigned to maintain the ingot solidification rate and the slag layer resistance at their programmed values. In addition, this system compensates for random transients in the high-voltage power supply, maintains a constant slag layer volume and chemical composition, and is designed for the use of a movable freezer mold.

The interaction of the system's lower level with the local control devices is provided by means of the interface shown in Fig. 6.3. The VTA-2000 terminal, used as the operating panel, is situated in the operator's workstation. A hard copy of the melting parameters is obtained on an alphanumeric printer.

The system's upper level is realized using a SM-4 computer, which is connected with each of the plant's furnaces by a data multiplexer that is based on an SM-1803 microprocessor. This multiplexer exchanges data between system levels and contains buffers for the temporary storage of data arrays.

The main task of the high-level system is to control the ESR process, based on an analysis of the thermal conditions of the casting and the slag layer, including calculations of temperature distribution in the casting and the slag. During melting, the calculated value of incremental power $\Delta P$ is transmitted as input data to the low-level system, where it is converted into a control action and then sent to the furnace actuators. In addition, the system's upper level performs various service functions, such as creating a melting report, compiling data on mill efficiency; scheduling the order of furnace operation; and choosing electrodes, printing documentation, and so on.

The ACS hardware is also organized for the fail–safe operation of the equipment. The system's lowest level is based on an ARShM-T regulator and associated local control devices; the next level is realised using an SM-1803 microprocessor and a complex of peripheral devices; and the uppermost level is comprised of an SM-4 computer. This type of system organization can be used for a wide variety of melting schedules, starting with simple individual control of each furnace up to simultaneous control of a group of furnaces, with optimal use of shop equipment and materials.

These descriptions of the systems show that various control systems for ESR furnaces can be realized by computer and digital control. The exact method chosen in each case depends on the melting technology in use at the plant, the equipment capacity, and, finally, on the quality requirements for the final product. In most cases, the most effective use of ACS is obtained only in conjunction with equipment that is modified to take advantage of automatic control and thus to provide opportunities for improvements in ESR technology, the development of new melting regimes, the production of high-quality steel, and reductions in cost.

## References

1   O.P. Bondarenko, E.D. Gladkii, and L.V. Chekotilo, "Modern control systems for ESR and their future development." Problems of Specialized Metallurgy 18. pp. 3–11 (1983)
2   "Electric and automatic equipment for electrothermal installations," Energiya, Moscow, p. 350 (1978)
3   A.A. Yakunin, G.N. Sidorin and V.M. Kochubeev, "SM-1803 microprocessor based hardware for automatic control systems for ESR," Metallurgical and Mining Industry 4, pp. 52–54 (1983)

# 7 Electroslag Remelting for the Production of Tool Steels

N.F. Yakovlev, Yu.M. Skrynchenko, S.I. Tishaev,
L.D. Moshkevich, A.N. Prohorov, and Yu.M. Politaev

The service characteristics of tool steels are largely determined by the grain size density and the homogeneity of the metallic matrix and the size, composition, and distribution of its nonmetallic inclusions. In many cases, conventional melting and casting methods cannot produce high-quality tool steel. The widespread use of electroslag remelting (ESR) has led to notable improvements in tool steel quality [1].

For 20 years the UkrNIIspetsstal, the E.O. Paton Electric Welding Institute, specialized metallurgical plants, and a number of manufacturing industry enterprises have worked together in order to improve ESR technology for tool steel production and to investigate the effect of ESR on tool steel properties. A detailed study on the development and introduction into mass production of high-speed and tool steel ingots and forgings has been made.

The results of this research show that ESR high-speed tool steels can be used directly in the as-melted state or with relatively little forging due to their isotropy. This simplifies tool manufacture because, in many instances, it is possible to omit the repeated forging of preforms.

Several complex technologies for the production of ESR high-speed steel with superior qualities were created and put into production at specialized metallurgical plants. These new technologies include such features as the annealing of cast or forged consumable electrodes and their subsequent ESR into ingots (0.7 to 4.0 tons) and SVTO combined with hammer and press forging [2]. The new processes can be used with such steels as R6M5-Sh, R6M5K5-Sh, and R6M5F3-Sh, first, in the production of 0.7-ton ingots in 300-mm diameter freezer molds using forged-alloy consumable electrodes (process A); second, in the production of 1.25-ton ingots in 425-mm diameter freezer molds (process B); and last, in the fabrication of 4-ton ingots, with a $500 \times 500$ mm cross section, by means of semicontinuous casting machines (SCCM) using permanent-mold cast electrodes (process C). The subsequent hot working of these ESR ingots either by hammer forging (porcesses A and B) or by SVTO and press forging (processes B and C) yields a high-quality steel, both micro- and macrostructurally, with practically no shrinkage porosity or ghost defects, in contrast to steel produced by regular electric arc melting. The macrostructure of R6M5-Sh steel produced using processes B and C has a carbide inhomogeneity in sections of 80 to 200 mm, which is consistently lower, by 1 to 2 units (Fig. 7.1), than that required by State Standards (GOST) 19625-73 for ESR forgings. The production

Author's affiliation: UkrNIIspetsstal, Zaporozhye, USSR.

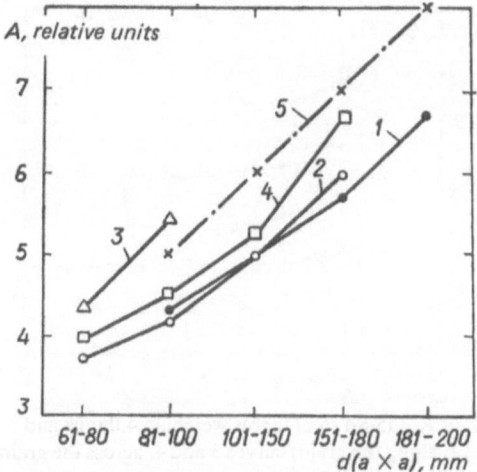

**Fig. 7.1.** Inhomogeneity of the carbide phase in R6M5-Sh steel forgings at a distance of a 0.5-mm radius, as a function of forging cross section. Curves 1, 2, and 3, ESR ingots weighing 4.0, 1.25, and 0.7 tons, respectively; curve 4, EA ingot weighing 1.0 ton; and curve 5, requirements of GOST 19265-73.

of rods over 100 mm in diameter from ESR ingots weighing 0.7 ton (process A) does not provide the necessary carbide homogeneity, due to the low amount of deformation.

It was determined that the use of ESR significantly widens the range of sizes and shapes of high-speed steel products and preserves the quality of semifinished products with diameters up to 320 mm. As an example, semifinished R6M5F3-Sh steel products as large as 320 mm in diameter produced form ESR ingots weighing 4 tons fully satisfy the requirements of GOST 19265-73 for high-quality ESR forgings with diameters of 180 to 200 mm.

The results of flexural strength tests on longitudinal and transverse samples of R6M5-Sh steel show the advantages of steel produced from 4-ton ingots, in comparison with ingots fabricated by process B (Fig. 7.2). This can be explained by the high amount of deformation of ingots from process B.

The results of commercial tests on ESR high-speed steel show that the wear resistance of R6M5F3-Sh milling cutters was increased by 30% to 40%, compared to cutters fabricated from EA steel (i.e., open-hearth electric arc steel). The wear resistance of large tools under impact loading increased by 50% to 80%.

It is therefore preferable to use high-speed ESR steels in the production of large, and complex cutting tools for service under conditions of high impact loading, due to their high tensile strength and homogeneous microstructure.

As a result of these service tests, it was possible to introduce more restrictive requirements for the micro- and macrostructure of ESR alloys (#3, #4 in GOST 19265-73), which are provided by the newly developed electroslag technologies (EST) for high-speed steel production.

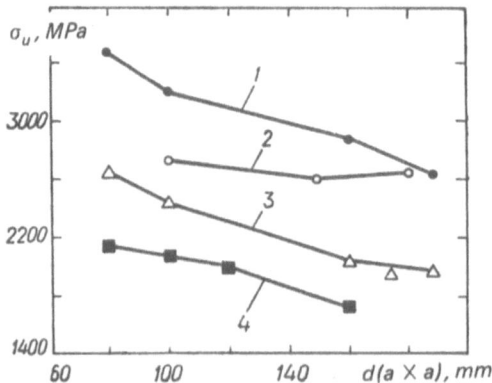

**Fig. 7.2.** Flexural strength of R6M5-Sh steel ingots weighing 4.0 tons and 1.25 tons, as a function of cross section. Curves 1 and 2, along the grain; curves 3 and 4, across the grain; curves 1 and 3, 4.0-ton ingot; and curves 2 and 4, 1.25-ton ingot.

The application of EST to the production of tool steels is another very promising development. Our investigations led to the development of technologies for the production of castings weighing 0.7 to 5.5 tons from tool steel brands 5HNM-Sh, 5H2MNF-Sh, 4H5MFS-Sh, 3H3MZF-Sh, 4H4VMFS-Sh, 5H3V3MFS, and others.

The structure and properties of ESR tool steels had increased density, homogeneity, toughness, and ductility compared to EA steels. The ESR tool steel also was more resistant to fatigue, especially in transverse sections from rods and other semifinished products with cross-sectional dimensions of $100 \times 250$ mm and greater. ESR tool steel is characterized by a lower anisotropy throughout the preform section. It was concluded that the effect of ESR on these characteristics of tool steels is similar to the effect of increased mechanical deformation.

The use of ESR steels led to the design of instruments with increased reliability and increased lifetime by factors of 1.3 to 2.0, as determined by numerous industrial test [3]. ESR most significantly affects the ductility and impact resistance determined at right angles to the grain structure of large billets (Table 7.1). As is clear from the table, the properties of samples of the two tool steels, 4H4VMFS-Sh and 3H3M3F-Sh, cut form the central part of semifinished products with cross-sectional dimensions of $180 \times 180$ mm, are improved 1.3 to 8.0 times, compared to those of EA steel, both at room temperature and at a working temperature of 600 °C. In addition, the anisotropy coefficient $n$ (according to Golikov) for 4H4VMFS-Sh steel at these two temperatures is reduced from 5.5 to 3.34 to 2.35 to 1.96, respectively, compared to steel manufactured by conventional melting technology; for 3H3M3F-Sh, $n$ is reduced from 2.68 to 1.52 to 1.64 to 1.39, respectively. This work and numerous commercial tests show that ESR tool steels are the steels of choice for the production of large or complex highly loaded dies for high-speed presses and forging hammers, as well as pres-

**Table 7.1.** Properties of ESR and EA tool steel preforms in the transverse direction

| Steel brand | Temperature (°C) | Properties | | | | | |
|---|---|---|---|---|---|---|---|
| 4H4VMFS | 20 | 1430/1320 | 1530/1410 | 8/1 | 14/3 | 10/5 | 2.53/5.50 |
| | 300 | 1210/1140 | 1330/1340 | 6/1 | 9/4 | 20/15 | 3.17/5.33 |
| | 600 | 740/700 | 910/850 | 12/9 | 32/14 | 17/20 | 1.96/3.34 |
| 3H3M3F | 20 | 1460/1350 | 1570/1450 | 10/4 | 18/9 | 22/18 | 1.64/2.68 |
| | 300 | 1220/1150 | 1380/1260 | 8/5 | 15/10 | 37/30 | 2.04/2.07 |
| | 600 | 810/770 | 980/900 | 14/11 | 41/36 | 37/27 | 1.39/1.52 |

Note: The numbers to the left of the slanted line refer to ESR steel; those to the right, to EA steel.

sure, die casting molds, which usually fail due to brittle fracture or fatigue fracture, or to erosion cracking.

A comparison of the properties of ESR and EA steels showed that 5HNM-Sh and 5H2MNF-Sh steel ingots are superior to forged EA steels (three to four forging passes) due to their high density and chemical and structural homogeneity and the absence of nonmetallic inclusions and dissolved gases.

Commercial tests showed that the wear resistance of large hammer and press dies (weighing 3 to 4 tons) fabricated from ESR steel of one of the above brands is 20% to 30% higher than that of forged EA steels of the same brand. Furthermore, ESR cast steel is more easily worked and tougher (especially in the transverse direction) compared to forgings, both at room temperature and at elevated testing temperatures; these improvements are most noticeable in smaller (up to 150 mm diameter), semifinished products, including those of the more highly alloyed 3H3VMF-Sh steel. The wear resistance of die components (sleeves, drawing dies, punches, etc.) fabricated from 3H3VMF-Sh steel ingots is 1.5 to 2.0 times higher than that of tools fabricated from EA billets. The use of ES ingots, and especially hollow ingots, reduces tool manufacturing time by 15% to 20% and increases the utilization factor of metal (UFM) up to 0.8 to 0.9.

ESR also favorably affects the structure and properties of steel for cold working, including the Lederburite steels H12M and H12F1. The use of an ES alloy, with its increased flexural strength and impact resistance, and its lower anisotropy, results in an improved tool reliability and a longer lifetime. ES steel can be successfully used in the production of complex tools for service under conditions of impact loading.

The structure and properties of bimetallic preforms produced in a newly designed installation for ES melting of cutting tools have been thoroughly studied. The cutter of such tools is fabricated of high-speed steel, whereas the shank is made of carbon steel. The consumable electrode was fabricated from worn-out tools of R6M5 steel, and type 45 steel was used for the head.

Further processing of preforms was accomplished in several different ways: isothermal annealing (850 °C, 1.5 hours; 740 °C, 4.0 hours); die forging of the preform (to reduce the cutter body to a diameter of 30 to 40 mm) and subsequent annealing; SVTO with or without subsequent forging; and further isother-

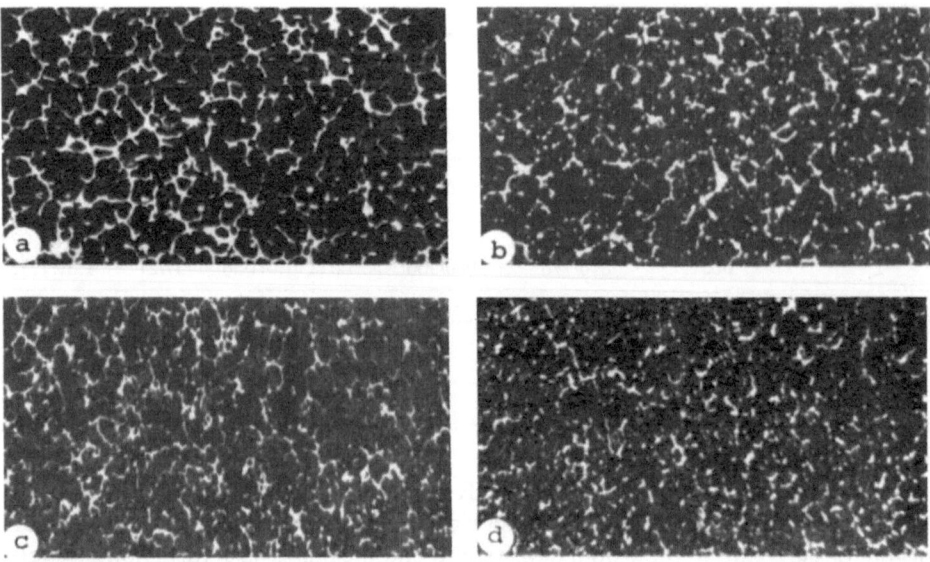

**Fig. 7.3.** Microstructure of the high-speed steel part of a bimetallic, semifinished product after various types of treatment: (a) conventional isothermal annealing; (b) forging plus annealing; (c) SVTO plus annealing; and (d) SVTO plus forging plus annealing.

mal annealing [2]. None of these methods produced observable defects at the junction of the two metals.

Analysis of the macrostructure of the high-speed steel part of the preform (Fig. 7.3) showed that, after melting and annealing, the carbide phase is present in the form of a fine eutectic carbide network (Fig. 7.3a). Forging stretches and partially ruptures this network (Fig. 7.3b). The SVTO, either by itself (Fig. 7.3c), or followed by forging (Fig. 7.3d), causes the carbide phase to coalesce and destroys the network structure. After coalescing, the carbide grains are oriented in the direction of the initial network.

Comparative analyses of commercial test results showed that the wear resistance of cutting tools cast from ESR alloys is comparable to that of forged EA alloy tools. The SVTO, and subsequent forging, increased the tensile strength of the semifinished product material from 140 to 160 to 290 to 324 MPa [4].

As a result of our investigations, it was concluded that tools with simple shapes that are to be used under normal cutting conditions should be fabricated from cast preforms that are conventionally annealed. For more complex tools, however, prior forging of the preforms with a small degree of reduction yields better results. The SVTO further improves the properties of the high-speed steel working part of the preform.

ESR of milling cutters was put into production at more than 10 manufacturing plants, which resulted in a net profit of over 1,200 rubles/ton of semifinished products.

An EST pioneered at the E.O. Paton Institute was subsequently developed in the UkrNIIspetsstal; it is a melting technique to produce hollow ES ingots and semifinished products of high-speed and tool steel. Hollow ES ingots are used for the fabrication of dies, large milling cutters, disk cutters for cold and hot machining, tube swaging dies, rolls for rolling mills, and so on.

Research on the technology and equipment for the production of hollow tool steel ingots was completed in the UkrNIIspetsstal [5]. The use of hollow ingots instead of solid ones yields a savings in material by reducing machining scrap by a factor of 1.5 to 2.0. It was found that this type of ingot can be used without being forged, or after forging, to replace EA alloy forgings. Moreover, the wear resistance of cast ES tools is of the same order or as much as four times the order of that of forgings [5]. Introduction of these new cutting tools significantly reduces cost to the amount of 1,000 rubles/ton of preforms, due to savings on tool steel, staff, machining time, and the more effective service of the equipment using the new tools. It should be pointed out that, in addition to being needed in the machine tool industry, hollow ES ingots are used in petrochemical, nuclear power, and other industries.

Shops producing hollow ES ingots and tool steel semifinished products have been built at several manufacturing plants. Creation of similar shops at ferrous metallurgy plants throughout the Soviet Union is feasible.

Thus, the use of EST in the production of tool steel can improve steel quality and supply industry with materials for highly efficient machine tools. In connection with this application, further increases in ESR tool steel production are planned in the near future.

# References

1   B.I. Medovar, G.A. Boiko, and S.P. Egorov, "The influence of ESR on improvement in service properties of 20HGS2H steel," in Remelting for Refining, Naukova dumka, Kiev, pp. 104–108 (1974)

2.  L.D. Moshkevich, V.F. Smolyakov, and T.I. Malinovskaya, "The influence of heating schedules on the properties of R6M5 steel," Tool and Bearing Steels, *1*, pp. 59–65 (1973)

3   S.I. Tishaev and Yu.M. Politaev, "The influence of ESR on tool steel quality," Steel *1*, pp. 31–38 (1978)

4   B.E. Natapov, Yu.N. Kuzmenko, and B.M. Nikitin, "Analysis of the structure and properties of tool preforms produced by means of ESR," Tool and Bearing Steels (1978)

5   A.K. Petrov, A.E. Koval, and K.K. Lyamtsev, "The quality of electroslag hollow ingots," in Electroslag Technology, Naukova dumka, Kiev, pp. 175–182 (1983)

# 8 Electroslag Remelting of Worn-Out Machine Tools

M.D. Tkachuk, V.I. Tashlykov, A.P. Stetsenko,
O.G. Seleverstov, G.A. Boyko, M.F. Zevakin,
V.V. Saranchuk, and L.E. Nebylitsin

Swaging die production is one of the main consumers of expensive and scarce tool steel. The production of sufficient quantities of tool steel for this purpose is very important. It is particularly important to improve the service characteristics and working life of the die and to reduce the manufacturing costs of tool production, all of which can be accomplished successfully through electroslag technology (EST).

A shop for recycling scrapped swaging dies by means of electroslag remelting (ESR) was built at the "Kovelselmash" plant in 1984 through the joint efforts of the E.O. Paton Electric Welding Institute and the RostNIITM Institute. This shop (Fig. 8.1) is situated in the forging shop and has a working area of 960 m². The shop is equipped with three USh-116 electroslag (ES) furnaces, a flux melter based on an A550U-12 apparatus, and an A550U-08 ES welder (ESW). All this equipment was constructed by the E.O. Paton Institute, except for the hardware of the USh-116 furnaces, which were produced at the "Kovelselmash" plant.

At this plant, dies are manufactured mainly from 5HNM steel. Large worn-out swaging dies are initially cut into pieces weighing up to 130 kg and are then hammer forged into electrode preforms of the required shape. The cross section of the forged preforms depends on the size of the ingots and on the requirements of the melting schedule. The five commonly used electrode cross sections are $100 \times 100$, $120 \times 130$, $140 \times 140$, $160 \times 160$, and $200 \times 200$ mm. The varieties of die preforms produced at the plant number in the tens. Small dies are reforged into sheetlike electrodes, rectangular, consumable electrodes with cross sections of $20 \times 70$, $20 \times 100$, $20 \times 120$, and $30 \times 160$ mm.

The consumable electrode is assembled from several reforged preforms that are welded together end-to-end. The length of the resulting electrode is determined by the weight of the ingot to be melted (Fig. 8.2). When the electrode body is assembled, a reusable head is welded on. Depending on the dimensions of the ingot to be produced, either a single electrode is remelted, or two to four electrodes are remelted simultaneously.

ES welding of consumable electrodes is performed in a dismountable, water cooled, copper permanent mold. Assembly and disassembly of the mold are fully mechanized: The mold bottom plate is raised and lowered by a hydraulic

Authors' affiliations: M.D. Tkachuk, V.I. Tashlykov, A.P. Stetsenko, and O.H. Seleverstov, the "Kovelselmash" plant, USSR: G.A. Boyko, M.F. Zevakin and V.V. Saranchuk; the E.O. Paton Electric Welding Institute, Academy of Science, UkSSR; and L.E. Nebylitsin, Rostov Manufacturing Technology Research (NII), USSR.

**Fig. 8.1.** Overall view of the ESR shop.

**Fig. 8.2** Consumable electrodes fabricated by welding reforged, worn-out dies.

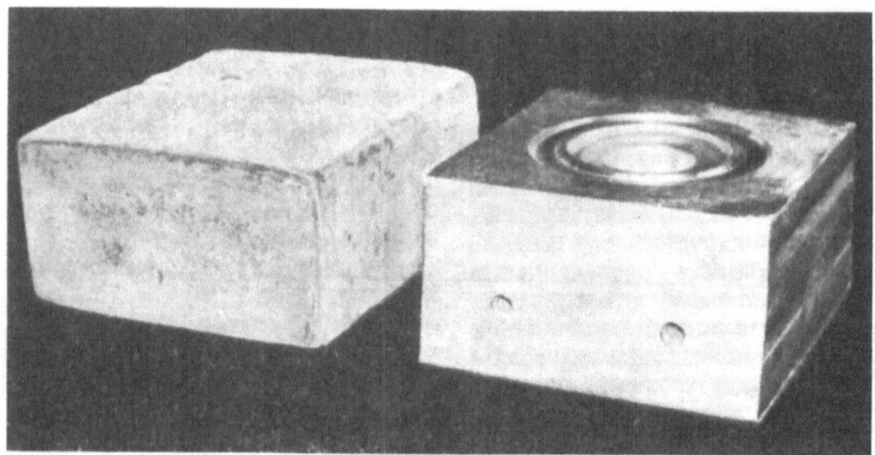

**Fig. 8.3.** Electroslag ingot and the final die product.

jack, and the side-forming panels are held in place by pneumatic actuators. Forged preforms are welded together in pairs in order to reduce the time it takes for the slag cap and the weld to cool. This reduces welding labor costs and increases the production rate.

Melting of die preforms is done in water-cooled copper freezer molds (Fig. 8.3), 1 to1.1 m in height. It yields ingots of sufficient size for the fabrication of two to three sets of die components. Round ingots are produced in closed shell copper freezer molds with water cooling.

Square preforms are melted in panel-type molds with slit cooling of the copper panels. The cooling water flows between a steel frame and the panel, which are separated by heat-resistant rubber seals. The special design of this freezer mold, in which two of the panels can be moved continuously between the stationary panels, makes it possible to melt ingots of various shapes and sizes using the same equipment. In all, only six round and six square universal freezer molds are needed to fabricate ingots to cover all commercially produced dies, which number over 200 types.

In order to reduce deformation of the panels and to increase panel lifetime, disk springs are used on the clamping bolts. The panel design permits repeated repairs by a combination of straightening, welding, and machining, which also prolongs the freezer mold's service life. The cross sections of ingots produced in panel freezer molds range from $120 \times 120$ to $320 \times 460$ mm; those produced in round freezer molds are from 230 and 380 mm in diameter.

Ingots are annealed in a furnace with a movable bottom plate. The ESR shop uses water from the mill's closed-loop water supply for cooling. The shop is also equipped with a self-contained, emergency water supply system, consisting of a pumping plant and a water reservoir.

The USh-116 installation use a liquid charge and can remelt consumable electrodes using a multielectrode bifilar circuit. Up to 30% of the flux is reused. For

**Fig. 8.4** Electroslag remelting of worn-out dies.

this purpose, the mill is equipped with an SMD-116 crusher for pulverizing the slag "plug" from ESR.

Swaging die preforms are cast with comparatively small machining allowances: 4 to 12 mm on the diameter or sides and 20 to 30 mm on the bottom and head parts (Fig. 8.4). The cropped heads and ends are not scrapped but rather are used as a seed charge. The high quality of the ES alloy increases the service life of forged dies by more than 1.2 to 1.5 over that of dies forged from conventionally melted alloy. Worn-out ES dies are reforged in a hot-stamping press into preforms for consumable electrodes; they are subsequently remelted. Thus, tool steel is continuously recycled. In recycling, certain precautions must be taken to prevent cross-contamination among different brands (grades, classes) of steel.

First, the incoming steel is inspected carefully. Before reforging, worn-out dies are stripped of components that are fabricated from other steels, such as knockouts, caps, and springs, as well as jammed forgings.

The possibility that other brands of steel may have been used for die manufacturing must also be considered. Thus, accurate marking, sorting, and tracking of the metal are required. ES ingots and tools from ESR steel, in addition to being marked with the steel brand, carry the abbreviation ESR and the number of the

melting run. When the ingot is cut into several pieces, these markings are transferred to each piece. Strict adherence to these rules prevents the intermixing
different steel brands during repeated reworking. Well-organized tracking of the
steel also permits better control of die tool wear resistance and the number of
remeltings the steel has undergone.

The introduction of ESR to reuse worn-out dies at the "Kovelselmash" plant
made it possible to satisfy 30% of plant's requirements for tool steel in just one
year. In the future, it is planned to increase this figure to 75% to 80%.

The reuse of worn-out dies through ESR yields a savings of 250 to 300 rubles/
ton of newly produced dies.

There are also provisions for recycling worn-out cutting tools fabricated from
R6M5 hgih-speed steel in the ESR shop. Damaged tools (mainly broaches and
drills), which earlier were scrapped, are now welding into consumable, one-
piece or divided electrodes. These electrodes are remelted in a small freezer mold
to form $120 \times 120 \times 350$-mm ingots. After annealing, an ingot is cut into templates of the required size for the tools to be fabricated. At present, research is
being carried out on remelting worn-out rolls of 4H8V2 steel. Future plans include recycling worn-out equipment of BrH bronze by means of permanent
mold ES casting (PMESC), which would significantly conserve nonferrous
alloys.

Thus, the introduction of EST at this plant has the following advantages: significant savings of expensive and scarce steels through recycling; reduction of the
manufacturing time of dies through decreasing machining allowances; increasing
die wear resistance, which gives additional savings in steel and labor costs; and
recycling tools and equipment made of the expensive alloys. In addition, small,
relatively compact ESR equipment can be simply deployed in the work area.

These data demonstrate the effectiveness of the use of EST in the manufacturing plant as a metal saving technology.

# 9 Electroslag Casting of Preforms for Forging Dies

G.A. Boyko, V.A. Miroshnichenko, B.K. Zykov, and S.M. Duplii

In the manufacturing industry, semifinished products for various components are generally produced by closed-die forging. In the manufacture of forging dies, increasing die wear resistance and decreasing machining allowances are the main ways of increasing the overall process efficiency, since they lead to reductions in manufacturing time and to savings on material costs. In the production of dies from open-hearth steel forgings, an excessive amount of the stock (40%) is removed by machining, due to large machining allowances. In addition, machining of the die cavity, the shank, and the lock expose the end-grain and reduce die strength in these areas.

Attempts to alleviate these problems by using cast dies made by pouring open-hearth steel into ceramic molds have not been very successful. The wear resistance of there cast dies is comparable to that of dies fabricated from rolled steel.

These drawbacks of dies made from forged or cast open-hearth steel, along with other drawbacks that show up in production and service, have been avoided at the Chernigov plant by the use of permanent mold electroslag casting (PMESC). This technique provides castings that are close in size and shape to the final die product and that significantly decrease machining allowances. In addition, the material has a high isotropy, homogeneity, and density, and is free of segregation defects. The wear resistance of dies is increased 1.3 times, on the average, compared to that of conventional forgings.

The production of forging die preforms by electroslag remelting (ESR) at the Chernigov plant was introduced in 1976 with the help of the E.O. Paton Electric Welding Institute. At that time, the mill equipment consisted of a U360B electroslag (ES) furnace and a U560 flux melting installation. A process for remelting worn-out forging dies that had been reworked on a forging press was developed. As a result, the problem of tool steel reuse at the plant was solved, and significant material savings were obtained. At present, there are no difficulties over tool steel availability in the forging die shop.

Following this successful trial with remelting tool steel scrap, the parent organization of the Chernigov plant, the MA "GAZ," decided to expand the electroslag technology (EST) shop in order to supply the whole association with steel for forged components. In 1984 three new ESR furnaces were introduced, which led to a fivefold increase in production capabilities (Fig. 9.1). Two ESR-2.5 L I1

Authors' affiliations: G.A. Boyko, E.O. Paton Electric Welding Institute, Academy of Sciences, UkSSR; V.A. Miroshnichenko, B.K. Zykov, and S.M. Duplii, the Chernigov Automobile Part Plant of the "GAZ" Manufacturing Association, USSR.

**Fig. 9.1.** Overall view of the electroslag technology shop at the Chernigov Automobile Part Plant.

furnaces were built to produce ingots weighing up to 2.5 tons. These furnaces are powered by EODTsN 4800/10-70U3, high-voltage transformers with a capacity of 2,500 KW-A, and a maximum current of 28 KA.

Cubical die preforms weighing up to 400 kg are produced in the ESR-0.25 VGL I1 frunace, which is equipped with a low-voltage, TPOV-630/100 VK transformer and which has a capacity of 630 KW-A and a maximum current of 10 kA.

Fomerly, all worn-out dies were reforged into shapes suitable for use in consumable electrode and were manually arc welded together to assemble the electrode. Now, a combined technology of electrode production is in use. Worn-out dies are directly welded into consumable electrodes in two A550U12 ESW devices that are powered by TShP 10/1 and TShS 3000/1 transformers. The electrodes for electroslag welding (ESW) consist of forged plates of the same composition as the remelted alloy, with a cross section of $20 \times 100$ mm and a length of 1 m (Fig. 9.2). Some of the worn-out dies (up to 30% of the total) are worked by drop forging to the required cross section and are then welded together in the semiautomatic A-765 welder to form rods; finally, a hot rolled rod is welded to the electrode head in the welder. Rolled stock is used for two reasons. First, the casting-to-product yield is 90%, and thus, there is a certain shortage of steel at the plant. Second, when the alloy is recycled, some elements are lost during the repeated melting. Close control of the steel's chemical composition is essential,

**Fig. 9.2.** Installation for electroslag welding of consumable electrodes from reforged, worn-out dies.

and to this end, additional alloying and deoxidizing, and adding "fresh" steel in the form of rolled stock, are necessary.

The two connected problems of losing steel elements and of maintaining the required chemical composition of the ingots have been carefully studied. It was found that repeated remelting of 5HNM tool steel produces voids in the casting due to carbon dioxide evolution during solidification. During remelting, up to 10% of the silicon present in the melt is lost, along with 5% of the manganese. The silicon, and to a lesser degree, the manganese present in the consumable electrode and in the flux act as residual reducing agents. The lowest silicon and manganese contents that will still prevent the development of gas voids in 5HNM steel remelted under an ANF-6 flux are 0.1% to 0.14% and 0.4%, respectively (State Standard GOST 5950-73 for this steel specifies a silicon content of 0.15% to 0.35%, and a manganese content of 0.5% to 0.8%). After worn-out forging dies made of this steel have undergone five remelting cycles, the silicon content has been reduced from 0.25% to 0.14%, which is close to the concentration at which gas pores develop. Gas pores are cone-shaped defects that start at the ingot bottom and extend some 200 to 300 mm to the surface.

Remelting of the steel under fluxes with an increased $SiO_2$ content (ANF-28 and ANF-29) did not have the desired effect—that the percent of silicon be maintained—and the use of these fluxes is extremely inefficient, due to high power consumption.

In order to restore the required chemical composition of repeatedly remelted alloy, a 60S2 steel rod containing up to 2% silicon was remelted with the main electrode. The results of chemical analysis after remelting of a consumable electrode of 5HNM steel with the addition of 5% 60S2 steel are given in Table 9.1. The concentration of silicon in the casting increases slightly, and the concentration of the other components remains about the same.

Good results were also obtained when aluminum was used as a reducing

**Table 9.1.** Chemical composition (%) of ingots produced by electroslag remelting with additional alloying durng remelting

| Studied object | C | Si | Mn | Cr | Ni | Mo |
|---|---|---|---|---|---|---|
| 5HNM (electrode) | 0.55 | 0.25 | 0.65 | 0.68 | 1.62 | 0.26 |
| 6OS2 (additive) | 0.6 | 1.80 | 0.75 | 0.22 | 0.20 | — |
| Casting | 0.54 | 0.28 | 0.6 | 0.66 | 1.56 | 0.25 |

**Table 9.2.** Longitudal variation of silicon concentration (%) in ingot from top (1) to bottom (4)

| Stduied object | 1 | 2 | 3 | 4 |
|---|---|---|---|---|
| Electrode | 0.12 | 0.12 | 0.10 | 0.09 |
| Casting | 0.18 | 0.17 | 0.15 | 0.15 |

agent. Here, an aluminum strip with a cross section of $50 \times 5$ mm was attached to the consumable electrode. Table 9.2 shows the results of chemical analysis after an electrode with a critical concentration of silicon was remelted. Remelting was done under 30 kg of ANF-6 flux, using 0.4 kg of aluminum for a 300-kg casting, at a current of 6 KA, and a voltage of 42 V. The numbers 1 through 4 in Table 9.2 correspond to different samples taken along the electrode and ingot from bottom to top. The ingot contains no pores, and the silicon concentration is restored, due to the flux.

For a number of years, ANF-6 flux has been reused because it is a relatively scarce, expensive material. The flux is melted in a U560 installation as follows: A fresh portion of flux is melted, after which is added up to 30% used flux. The slag "plugs" that are to be reused are crushed into pieces up to 30 mm in size in a

**Fig. 9.3.** Freezer mold with a two-directional, adjustable, cross-sectional melting zone.

2-ton M1343 hammer mill. The savings from reusing flux amount to 9 rubles/ton of remelted steel.

All the ESR furnaces in the shop use a liquid charge. The slag is siphoned through a groove in the bottom plate, which is a more efficient process compared to processes in which slag is poured from above or processes in which a solid charge is used.

In order to use less tool steel and to increase die manufacturability, freezer molds are provided with water-cooled copper inserts or machined grooves in the panels in order to obtain castings with dimensions that are close to those desired for die shanks and locks.

Freezer molds are designed to allow continuous adjustment of the transverse cross section of the melting zone in two directions (Fig. 9.3) to produce die castings with a minimum machining allowance. This equipment gives the shop the capability of producing over 60 different types of forging dies.

In addition to achieving the designed output of the EST shop, other work is being carried out to improve ingot quality.

# 10 Electroslag Technology at the "Izhstal" Manufacturing Association

N.A. Ponamarev, K.K. Zhdanovich, N.P. Trebov,
E.A. Upshinski, M.K. Zakamarkin, M.A. Loiferman,
and M.M. Lipovetski

The electroslag remelting (ESR) shop of the MA "Izhstal" was set up 25 years ago. At that time, there was only one R-909 furnace. Later, two additional R-951 furnaces were installed. At the beginning of the 1970s, a new ESR shop was built, with OKB-1155A and OKB-1155B type furnaces, and it attained its designed output in 1977. At that time, the production of square ESR tool steel ingots with a cross section of 485 × 485 mm was characterized by the following coefficients: productivity of the remelting process, 500 kg/hour; specific power consumption, 1400KW-h/t; and material consumption coefficient, about 1.65 (ingot weight, 1350 kg; final product yield, 820 kg). In view of these figures, most of the development efforts at the MA were directed toward improving the efficiency of the process (reducing material and electric power consumption and increasing furnace productivity).

The final product yield was increased by reducing the length of crop ends. The existing process called for cropping 7% of the ingot length to remove the shrinkage pipe. However, this does not always ensure that slag from the pipe would not penetrate the casting head. The depth and size of the slag pipe depends on the shape and volume of the molten metal pool. In order to avoid pipe formation, it is necessary either to decrease molten steel volume or to prevent the development of a "bridge" in the upper part of the metal pool after the remelting process is terminated. There are many well-known methods to reduce the volume of the molten steel, but they are very complex, in practice, or so time consuming as to be uneconomical. A simpler method turned out to be to heat the ingot head after termination of remelting. In numerous experiments, it was found that FS-75 ferrosilicon, a good heating material, reduced the ingot crop heads to 5%.

It is well known that increasing the solidification rate increases furnace productivity and slag reactivity and reduces specific power consumption. However, an increased solidification rate reduces the time available for refining reactions between the molten steel and the slag to occur and increases molten metal volume. In the course of investigations carried out to improve the electroslag (ES) process, various steels were melted with solidification rates from 500 to 1,500 kg/h. The resulting ingots had optimal electrical parameters, which permitted solidification rates as high as 800 to 1000 kg/h, depending on steel type. The ESR process is presently computer controlled and routinely yields high-quality steel.

---

Authors' affiliations: "Izhstal" Manufacturing Association, USSR.

In order to prevent erosion of the freezer mold bottom plate during the initiation of remelting and to provide good electrical contact between the ingot and the bottom plate, special seed charges were fabricated from sheet 20 to 50 mm thick. In addition, a new charging technique that used sectional washer seed charges was introduced. This made it possible to reduce the consumption of charged steel and to decrease ingot bottom trimmings to 2%. Surface conditioning of consumable electrodes is eliminated due to the use of electrodes with a closed shrinkage pipe produced by semicontinuous casting.

These new developments further improvemed the ESR characteristics: average productivity, 900 kg/h; specific power consumption, 1,150 kW-h/t; material consumption coefficient, about 1.24 (ingot weight, 1,065 kg; final product yield, 860 kg).

In recent years, the most important research has centered on developing methods to obtain solid ingots with a diameter, of up to 500 mm in stationary freezer molds and ingots of varied shapes and dimensions in movable freezer molds.

Low-alloy constructional steel ingots were produced using an electrode-bottom plate and several different processes, which are listed in Table 10.1. Data from tests on the properties of the steel are shown in Table 10.2 and 10.3, based on the gas content, shown in Table 10.4, and on the sulfide contamination, shown in Table 10.5. The quality of the ingots was checked by magnaglo and ultrasonic inspection; internal flaws were detected in many ingots (Fig. 10.1).

To discover the reason why internal flaws develop, ultrasonic inspection was performed on a number of ingots that were melted at various solidification rates. The averaged inspection results showed a direct dependence of defect length on soildification rate: For an average solidification rate greater than 900 kg/h, internal defects were observed to extend throughout the ingot length. Minimum defects were obtained at a solidification rate of 650 kg/h. No defects were observed for solidification rates of 600 kg/h or lower. Internal flaws occur in the form of fracture seams that develop during solidification of the molten steel. A series of ingots that were produced using the same electrical parameters had a defect-free structure. Subsequent ultrasonic inspection of this batch of ingots showed no internal flaws.

Low-cost cast dies for hot forging, fabricated by ESR and having a higher wear resistance than forged steel dies, have been successfully introduced in the industry. When the wide range of forging dies that are in use was taken into account, it was found to be necessary to design a freezer mold that would be suit-

**Table 10.1.** Different ESR schedules for OKB-1155A furnaces

| Schedule | Flux type | Flux consumption (kg/t) | Average solidification rate (kg/hour) |
|---|---|---|---|
| I | ANF-32 | 22 | 400 |
| II | ANF-6 | 24 | 400 |
| III | ANF-6 | 24 | 800 |

**Table 10.2.** Mechanical properties of ESR, low-alloy, constructional steel

|  | The location of cross sectional specimens | | | | | | | | | | | |
|---|---|---|---|---|---|---|---|---|---|---|---|---|
|  | I | | | | | | II | | | | | | III | | | | | |
|  | Edge | | | 1/2 Radius | | | Edge | | | 1/2 Radius | | | Edge | | | 1/2 Radius | | |
| The location of specimens cut along the casting height | $\sigma_{0.2}$ (MPa) | $\sigma_v$ (MPa) | $\psi$ (%) | $\sigma_{0.2}$ (MPa) | $\sigma_v$ (MPa) | $\psi$ (%) | $\sigma_{0.2}$ (MPa) | $\sigma_v$ (MPa) | $\psi$ (%) | $\sigma_{0.2}$ (MPa) | $\sigma_v$ (MPa) | $\psi$ (%) | $\sigma_{0.2}$ (MPa) | $\sigma_v$ (MPa) | $\psi$ (%) | $\sigma_{0.2}$ (MPa) | $\sigma_v$ (MPa) | $\psi$ (%) |
| Top | 1,270 / 1,270 | 1,400 / 1,380 | 49 / 44 | 1,280 / 1,280 | 1,400 / 1,390 | 26 / 35 | 1,280 / 1,280 | 1,400 / 1,410 | 51 / 51 | 1,300 / 1,270 | 1,420 / 1,400 | 32 / 38 | 1,270 / 1,200 | 1,410 / 1,400 | 52 / 43 | 1,280 / 1,260 | 1,400 / 1,400 | 27 / 28 |
| Center | 1,280 / 1,280 | 1,400 / 1,400 | 48 / 46 | 1,250 / 1,270 | 1,380 / 1,400 | 25 / 30 | 1,300 / 1,280 | 1,430 / 1,400 | 48 / 38 | 1,280 / 1,290 | 1,400 / 1,420 | 29 / 30 | 1,200 / 1,270 | 1,420 / 1,400 | 51 / 52 | 1,270 / 1,270 | 1,400 / 1,400 | 31 / 24 |
| Bottom | 1,270 / 1,260 | 1,400 / 1,400 | 51 / 24 | 1,270 / 1,260 | 1,400 / 1,400 | 33 / 17 | 1,280 / 1,270 | 1,410 / 1,400 | 51 / 46 | 1,280 / 1,260 | 1,380 / 1,390 | 32 / 33 | 1,270 / 1,270 | 1,400 / 1,400 | 39 / 38 | 1,280 / 1,270 | 1,400 / 1,400 | 12 / 27 |
| State Standard #V5182-78 on the forgings is no less than | 1,200 | — | 20 | 1,200 | — | 20 | 1,200 | — | 20 | 1,200 | — | 20 | 1,200 | — | 20 | 1,200 | — | 20 |

Note: I to III, ESR schedule number; figures above the line are steel properties in the longitudinal direction; figures under the line are properties in the transverse direction.

**Table 10.3.** Impact resistance of ESR, low-alloy, constructional steel

| The location of specimens cut along the casting height | The location of cross-sectional specimens | | | | | | | | | | | |
| | Edge | | 1/2 Radius | | Edge | | 1/2 Radius | | Edge | | 1/2 Radius | |
| | +20 | −50 | +20 | −50 | +20 | −50 | +20 | −50 | +20 | −50 | +20 | −50 |
| | I | | | | II | | | | III | | | |
| Top | 4.0 / 3.9 | 3.8 / 3.8 | 3.2 / 3.1 | 3.1 / 3.1 | 3.6 / 3.0 | 3.6 / 2.3 | 3.6 / 2.7 | 3.3 / 1.9 | 3.6 / 3.5 | 3.4 / 2.8 | 4.2 / 3.0 | 3.2 / 2.1 |
| Center | 4.1 / 4.1 | 3.2 / 3.8 | 3.2 / 3.3 | 3.0 / 3.2 | 4.9 / 3.9 | 3.9 / 3.2 | 3.8 / 3.2 | 2.9 / 2.9 | 4.0 / 2.8 | 3.1 / 2.7 | 4.0 / 2.9 | 3.1 / 2.5 |
| Bottom | 4.8 / 3.5 | 3.6 / 2.8 | 4.1 / 3.2 | 3.5 / 2.5 | 5.4 / 3.3 | 4.6 / 3.7 | 4.1 / 3.0 | 3.5 / 2.2 | 3.8 / 3.9 | 3.2 / 3.8 | 3.0 / 2.8 | 3.0 / 3.6 |

Note: I to III, ESR schedule number; figures above the line are steel properties in the longitudinal direction; figures under the line are properties in the transverse direction; +20 and −50, test temperature (°C).

able for dies of various shapes and sizes. This problem was solved by constructing a universal freezer mold with interchangeable panels that could be used to melt ingots of various lengths by using the appropriate height panels.

It is relatively straightforward to obtain ingots with a high-quality surface finish in stationary freezer molds. However, in installations with movable molds in which there is an opposing motion of the electrode and the mold, obtaining ingots with a satisfactory surface finish is an ongoing problem. Choosing the correct flux composition for ESR in a movable freezer mold is of great importance. The most useful fluxes for this purpose are ones that contain $SiO_2$, such as

**Table 10.4.** Oxygen content and contamination by oxygen-containing nonmetallic inclusions of ESR, low-alloy, constructional steel (%)

| The location of specimens cut along the casting height | The location of cross-sectional specimens | | | | | | | | |
| | Edge | 1/2 Radius | Center | Edge | 1/2 Radius | Center | Edge | 1/2 Radius | Center |
| | I | | | II | | | III | | |
| Top | 0.0071 / 0.034 | 0.0075 / 0.042 | 0.0064 / 0.038 | 0.0046 / 0.018 | 0.0047 / 0.028 | 0.0042 / 0.027 | 0.0069 / 0.023 | 0.0064 / 0.034 | 0.0071 / 0.030 |
| Center | 0.0046 / 0.010 | 0.0046 / 0.023 | 0.0061 / 0.016 | 0.0033 / 0.025 | 0.0036 / 0.008 | 0.0029 / 0.008 | 0.0051 / 0.014 | 0.0053 / 0.025 | 0.0056 / 0.025 |
| Bottom | 0.0028 / 0.013 | 0.0026 / 0.008 | 0.0027 / 0.013 | 0.0031 / 0.015 | 0.0028 / 0.006 | 0.0030 / 0.011 | 0.0039 / 0.022 | 0.0037 / 0.014 | 0.0040 / 0.017 |

Note: Figures above the line indicate oxygen content; figures under the line indicate nonmetallic inclusion content; I to III, ESR schedule number.

**Table 10.5.** Contamination of ESR, low-alloy, constructional steel by sulfides

| The location of specimens cut along the casting height | The location of cross-sectional specimens | | | | | | | | |
|---|---|---|---|---|---|---|---|---|---|
| | Edge | 1/2 Radius | Center | Edge | 1/2 Radius | Center | Edge | 1/2 Radius | Center |
| | I | | | II | | | III | | |
| Top | 0.0011 / 21 | 0.005 / 35 | 0.0014 / 89 | 0.001 / 20 | 0.0029 / 21 | 0.0019 / 21 | 0.0012 / 13 | 0.0015 / 14 | 0.0093 / 67 |
| Center | 0.0019 / 14 | 0.0028 / 16 | 0.0022 / 27 | 0.0026 / 39 | 0.001 / 17 | 0.0008 / 14 | 0.0008 / 11 | 0.0032 / 20 | 0.003 / 35 |
| Bottom | 0.0002 / 6 | 0.0005 / 13 | 0.0014 / 25 | 0.0005 / 9 | 0.0007 / 10 | 0.0018 / 16 | 0.0016 / 27 | 0.0012 / 12 | 0.004 / 19 |

Note: Figures above the line indicate inclusion content (%); figures under the line indicate the number of inclusions; I to III, ESR schedule number.

ANF-28, ANF-29, and ANF-32, which were developed in the E.O. Paton Electric Welding Institute.

The $CaF_2$–$Al_2O_3$ system was chosen for developing new flux compositions for use in movable freezer molds. To reduce the melting point and to extend the solidification range, $SiO_2$, was added; $BaCl_2$ was added to retain refining properties. Slag that contains 5% to 15% $SiO_2$, 5% to 15% $BaCl_2$, 40% to 60% $CaF_2$, and 20% to 25% $Al_2O_3$ has a melting point between 1,100 and 1,290 °C, an electrical conductivity of 2.0 to 4.0 ohm-1, and a viscosity of 0.2 to 0.4 Pa-s.

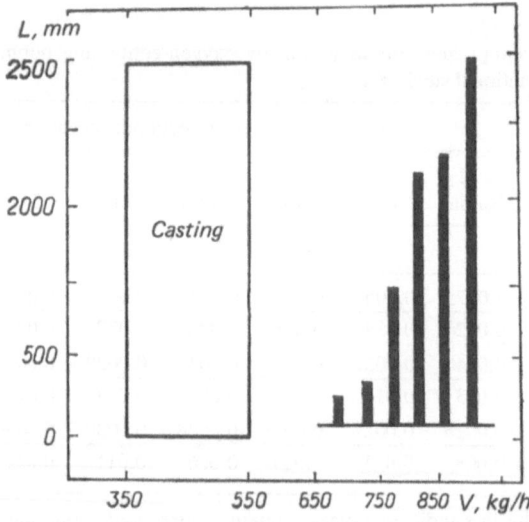

**Fig. 10.1.** Dependence of the length $L$ of axial defects on the ingot solidification rate $V$.

**Table 10.6.** Properties of ESR tool steels

| Steel type and object studied | $\sigma_v$ (MPa) | $\sigma_{0.2}$ (MPa) | $\delta$ (%) | $\psi$ (%) | KCV (J/m²) | KCT (J/m²) | HRC |
|---|---|---|---|---|---|---|---|
| 5HNM-Sh, casting | 1,449 | 1,357 | 9.6 | 40.5 | 38.7 | 13.0 | 42.5 |
|  | 1,449 | 1,330 | 8.6 | 37.9 | 31.4 | 10.9 | 42.0 |
| 5HAMNF-Sh, casting | 1,602 | 1,489 | 9.9 | 37.9 | 37.5 | 9.1 | 45.5 |
|  | 1,594 | 1,493 | 8.2 | 26.6 | 37.0 | 9.6 | 45.5 |
| 5HM, forging | 1,520 | 1,450 | 8.0 | 19.0 | 33.0 | — | 39.0 |
|  | 1,500 | 1,430 | 4.0 | 11.5 | 30.0 |  | 39.0 |

Note: Figures above the line, longitudinal samples; figures under the line, transverse samples.

These characteristics satisfy the requirements for ESR fluxes. For commercial use, a flux containing 80% ANF-6, 15% $SiO_2$, and 5% $BaCl_2$ was proposed. This flux is used in the production of various types of cast forging dies and provides a satisfactory surface finish.

At present, cast forging dies are produced from 5HNM and 5HAMNF steels for drop forging presses with hammers weighing from 1 to 5 tons and for presses with capacities up to 6.3 MN.

The properties of several brands of ESR tool steel are shown in Table 10.6. These data once again confirm that the properties of ESR alloys are almost identical to those of forged, open-hearth steels. The ESR ingots have a higher isotropy than the forgings have. The final heat treatment of cast forging dies of 5HNM-Sh steel was performed using the same schedule as that used for dies forged from open-hearth steel. Commercial tests of ESR cast steel dies show an increase in service life from 1.1 to 2.0.

The results presented in this chapter show the effectiveness of using ESR to improve the service characteristics of steels.

# 11 Electroslag Remelting of Manganese

Yu.V. Latash, V.A. Yakovenko, I.Yu. Lyuty, E.V. Butski,
S.V. Bogdanov, and V.P. Kubikov

The achievable level of service characteristics of manganese-based microalloyed steels is determined mainly by their purity. The Mr-O brand of electrolytic manganese that is used for these alloys requires additional refining due to its high content of sulfur and dissolved gases. Additional refining is usually performed by radiofrequency induction (RFI) melting under a slag layer. In order to use this method, special melting techniques had to be developed because of some anomalies of the refining process. For example, exceptionally severe conditions require the use of special, high-quality refractories, for the furnace and the crucible linings. Relatively low slag temperature reduces slag reactivity and necessitates refining times of six to eight hours, during which time repeated adjustments of the slag composition are required, along with running-off of the slag. All these requirements make the process cumbersome and lead to significant losses of manganese [1, 2].

Accordingly, a new method of refining electrolytic manganese, based on electroslag remelting (ESR), was developed at the E.O. Paton Electric Welding Institute. The new method depends on the sequential fusion of flakes of electrolytic manganese in a slag layer that acts as both a heat source and a refining medium. Current is supplied to the slag layer by permanent electrodes [3].

The use of permanent electrodes for ESR of manganese makes it possible to vary slag layer voltage, temperature, composition, and thickness. Due to the high slag temperature and the large surface area of the manganese flakes (300 m/t), melting occurs very rapidly, practically within the time it takes the flakes to pass through the slag layer. Furthermore, the conditions required for manganese refining and for oxide reduction are provided simultaneously. All of these factors result in a high degree of refining and a high process productivity.

The authors have studied the dependence of the degree of manganese desulfurization on the composition of the slag, the furnace atmosphere, and other process parameters. In addition, thermal and energy balances and the efficiency of the process were determined.

An analysis of manganese desulfurization revealed that the effect of the slag composition is the dominant effect. The best results were achieved using slags based on a $CaF_2-CaO-Al_2O_3$ system and were characterized by a high equilibrium sulfur distribution coefficient, $L_s = 30$ to $40$. Numerical methods were used

---

Authors' affiliations: Yu.V. Latosh, V.A. Yakovenko, and I.Yu. Lyuty, E.O. Paton Electric Welding Institute, Academy of Sciences, UkSSR; E.V. Butski, S.V. Bogdanov, and V.P. Kubikov, "Electrostal" plant, USSR.

to obtain regression equations from the experimental data, which showed the relation between parameters describing manganese quality and slag composition. These regression equations are expressed as a power series as follows:

$$[S] = 0.05628 - 0.0494A - 0.0739B - 0.0193C + 0.0231A + 0.0225AB$$
$$- 0.0175AC + 0.0232BC \ldots$$
$$[Si] = 0.23 + 5.073B + 1.469C - 16.631B + 1.575AC \ldots,$$

where $A = Mslag/MMn$; $B = MCaO/Manf\text{-}6$; $C = MSiO_2/Manf\text{-}6$; and MMn, Mslag, MCaO, $MSiO_2$, and Manf-6 stand for the weights of Mn, slag, CaO, $SiO_2$, and ANF-6 flux, respectively.

For convenience, the desulfurization studies results are presented graphically in Fig. 11.1. The degree of manganese desulfurization is determined mainly by the volume and the basicity of the slag. The addition of $SiO_2$ to the slag suppresses desulfurization.

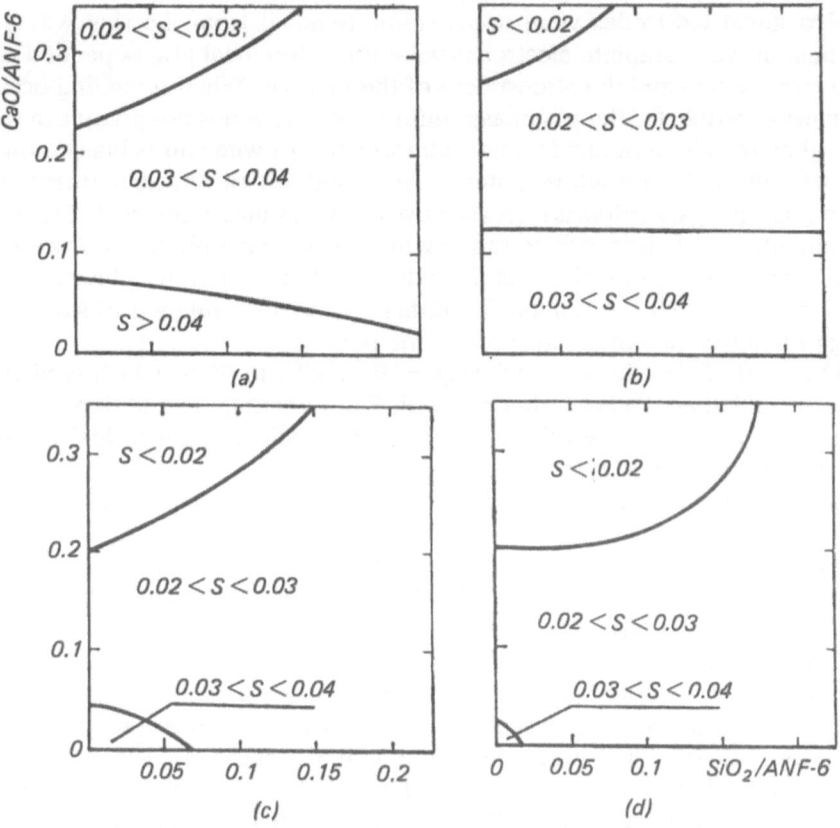

**Fig. 11.1.** Dependence of sulfur concentration in ESR manganese on the concentration (by weight) of CaO and $SiO_2$ in an ANF-6 flux, for manganese-to-slag mass ratios of (a) 0.2; (b) 0.5; (c) 0.75; and (d) 1.0.

The degree of desulfurization is significantly affected by the presence of MnO in the slag. In the course of remelting, the concentration of MnO increases and can reach 20%. This increase is caused by the transport of manganese oxides from the charge to the slag and by the reaction of manganese with atmospheric oxygen, which is transported through the slag due to the presence of multivalent oxides of manganese. The net effect is to increase the viscosity and melting point of the slag and to reduce desulfurization.

In some experiments, graphite electrodes were used to supply power to the slag layer. It was observed that the manganese oxide content in the slag was reduced to 0.5%, due to reduction by carbon from the electrodes; furthermore, the sulfur content of both the slag and the manganese was reduced to trace amounts, even though the carbon content of the manganese increased to 1%. The fact that the oxygen content of the slag significantly affects the degree of desulfurization made it necessary to modify the refining method. To that purpose, calcium was added to the slag, which increased the degree of desulfurization by 10% under equivalent remelting conditions.

Sulfur is removed from the slag into the gas phase. It was observed that when cooled metal electrodes were used, sulfur removal from the slag was not as efficient as when graphite electrodes were used. Removal also depended on the slag composition and the atmosphere of the furnace. When remelting occurs in an inert atmosphere, the gas phase sulfur concentration is not greater than 5%; but when remelting occurs in an air atmosphere, or when air is blasted through the slag, the gas phase sulfur content is increased to 10% and 20%, respectively. However, manganese loss is increased two or three times in this case (Fig. 11.2b). The addition of caustic soda to the slag increases the gas phase sulfur content to 60%, probably because the volatile sodium sulfide evaporates. Thus, desulfurization during ESR is accompanied either by the accumulation of sulfur in the slag or by formation of gaseous sulfur compounds.

The level of desulfurization during ESR is 85% to 95%, which is higher by 10% to 15% than the level obtained with RFI remelting; in addition, the desulfurization rate is increased three to four times, while manganese loss is reduced up to five times (Fig. 11.2).

The electrolytic technique yields manganese metal with an increased dissolved gas content of up to 800 cm$^3$/100 g of manganese, including up to 0.2% oxygen, 0.1% hydrogen, and 0.1% nitrogen. After ESR, the gas content is reduced by 80 to 100 cm$^3$/100g manganese (0.01% to 0.02% $O_2$, 0.003% to 0.005% $H_2$, and 0.004% to 0.006% $N_2$).

Carbon is another impurity that has an adverse effect an the quality of micro-alloyed steels. As noted above, graphite electrodes raise the carbon content of remelted manganese to unacceptable levels. When cooled metal electrodes are used for remelting, the slag is the only source of manganese. When commercial fluxes containing no more than 0.03% of carbon are used, the carbon concentration in the refined manganese is 0.04% to 0.06%; that is, it remains at the original level. When remelting is performed with a liquid charge using graphite electrodes and a graphite crucible, only insignificant increases in carbon content are

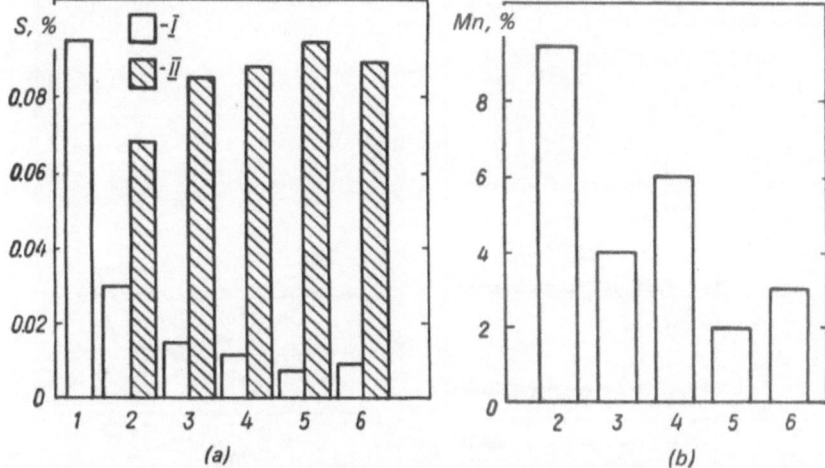

**Fig. 11.2.** (a) Sulfur concentration (*I*) and level of desulfurization (*II*); (b) loss of manganese during ESR. (1) Unrefined manganese charge; (2) radiofrequency induction melting; (3) ESR in the air; (4) air blasted through slag layer during ESR; (5) ESR with added reducing agents; and (6) ESR in an inert atmosphere.

observed in the bottom part of the ingot. Experimental data show that the silicon content of the manganese increases with increasing silicon oxide concentrations in the slag, with the slag volume, and with decreasing calcium oxide concentrations. Our experiments showed the regulating effect of manganese oxide on the reduction of silicon from the silicon oxide in the slag. The direction of this reaction is determined by the concentration and activity of manganese oxide and silicon oxide in the slag. When using commercial slags containing up to 2.5% $SiO_2$, it is necessary to keep the manganese oxide concentration in the slag between 8% and 10% to prevent the incorporation of silicon into the refined manganese.

Contamination of manganese by metal from the permanent electrode was not detected. The design of these cooled metal electrodes is such that it is possible to use remelting regimes wherein a thin layer of slag (skin) solidifies on the electrode surface, which prevents the electrode from coming into contact with the molten slag and thus eliminates slag contamination by the electrode material and reduces electrode wear.

ESR manganese ingots are smooth and are easily separated from the solidified slag after cooling. These ingots are characterized by a uniform distribution of impurities lengthwise as well as radially (Fig. 11.3). Differences in sulfur concentration at the head and tail ends of the ingot are caused by sulfur accumulation in the slag in the course of remelting and, thus, a reduction in refining ability. In contrast, increased concentrations of carbon, silicon, and iron in the tail end are caused by the incorporation of these components from the slag during the initial stages of remelting. Variations in sulfur content along the diameter

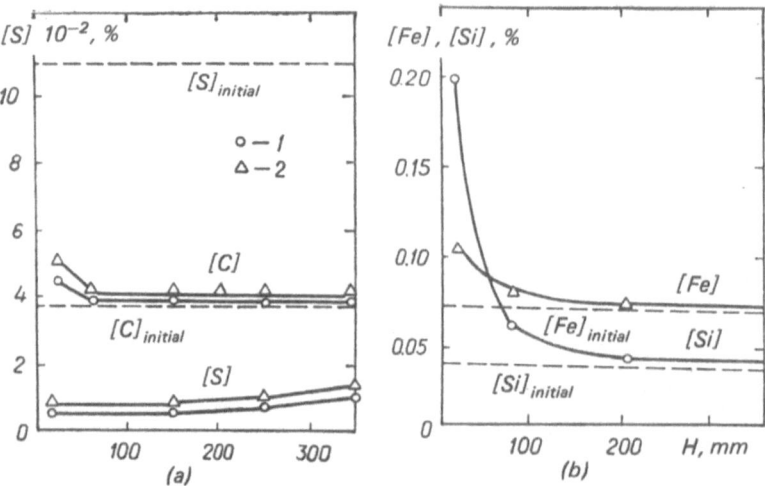

**Fig. 11.3.** Distribution of sulfur and carbon (a) and iron and silicon (b) in the interior of ESR manganese ingots (1) and on their surface (2).

are no greater than 10% and are observed mainly close to the surface of the ingot. The characteristics of microalloyed steels that are produced by ESR manganese satisfy the quality requirements of State Standard (GOST).

Analysis of the energy balance of ESR of manganese reveals that 15% to 25% of the input energy is consumed in heating and melting the metal. Heat losses through the cooled metal electrodes account for 10% to 25% of the energy consumption, depending on such variables as slag composition, applied voltage, and the power supply circuit. The specific power consumption varies between 1,500 and 2,000 kW-h/ton of remelted metal.

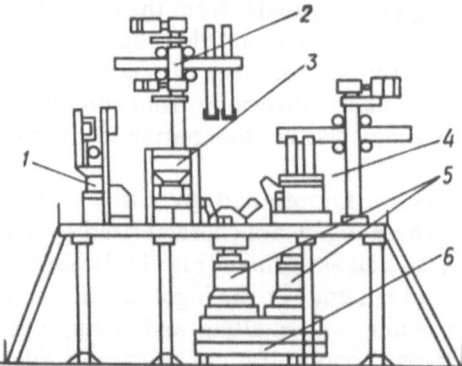

**Fig. 11.4.** Schematic diagram of the UO-105 installation for ESR of manganese: (1) hydraulic unit; (2) electrode lifting mechanisms with cooled metal electrodes; (3) hopper with vibrating feeder; (4) flux melting furnace; (5) freezer molds; and (6) transfer car.

Based on the results of these investigations, a commercial installation was developed for remelting and refining steel and other alloy charges in friable and in lump form (Fig. 11.4). The ESR installation is an electroslag furnace with 400-mm-diameter freezer molds and a furnace for initial flux melting. It is equipped with hoppers and vibrating feeders for the storage and the gradual addition of slag and manganese. Graphite electrodes are used for melting slag in the flux melting furnace, whereas cooled permanent metal electrodes are used for remelting manganese metal in a freezer mold. The bimetallic working ends of the metal electrodes consist of an outer covering made of a high melting point alloy, which is resistant to erosion by the slag, and an inner core with a high thermal conductivity. The weight of manganese ingots produced in this installation is 500 kg; the solidification rate is 300 to 500 kg/h.

# References

1   K.P. Bikezin, M.M. Klyuev, and S.V. Ilyin, "Studies of manganese desulfurization by synthetic slags." Manganese Mining, Enrichment and Processing *39*, pp. 82–92 (1974)
2   Yu. A. Gratsianov, B.N. Putimtsev, and V.V. Molotilov, "Precision alloy metallurgy," in Metallurgiya, Moscow, 1975, p. 488 (1975)
3   I.Yu. Lyuty and Yu.V. Latash, "Electroslag melting and refining of alloys," Naukova dumka, Kiev p. 188 (1982)

# 12 Electroslag Remelting of Aluminum Bronzes Exhibiting the Memory Effect

V.K. Larin and D.F. Chernega

The term *memory effect* (*ME*) refers to the ability of compounds that are fabricated by plastic deformation of certain materials to regain their initial shape following heating. This effect is more fully seen in alloys that have undergone a thermoelastic martensitic transformation. In all of the applications of alloys exhibiting ME (engines and drives, thermomechanical joints, temperature-sensitive elements, and so on), it is essential that they regain their shape under sharply defined thermal conditions, which are determined by service conditions. The thermal conditions are functions of the composition and the stress state of the material, but are determined primarily by the alloying element content. According to data obtained by the present authors, each 0.1% (by weight) of aluminum in Al–Cu–Mn alloys decreases the characteristic temperature of the martensitic transformation, on average 15 °C. This fact is of prime importance and should be taken into account when the possibility of electroslag remelting (ESR) of alloys with ME is considered.

The necessity of using ESR processing of alloys exhibiting ME is dictated by the strict quality requirements for the alloy, which are determined by subsequent processing steps and by service conditions. Semifinished products of various shapes, such as strip, rod, tube, and wire, are fabricated from ESR ingots by rolling or hot pressing (with additional extrusion, if necessary). These products are used to fabricate various temperature-sensitive elements (springs, sleeves, etc.) that will undergo repeated heating–cooling cycles and alternating strain and stress. In this type of service, an alloy with a compact grain structure, a high chemical homogeneity, and minimal amounts of gas and nonmetallic inclusions is required.

Electroslag-remelted alloys satisfy all the above requirements, especially in ingots obtained by remelting high thermal conductivity alloys and copper alloys. The high thermal conductivity permits a large temperature gradient to occur in front of the solidification front, which, under conditions of uniform heat flows (as occurs in copper water-cooled freezer molds), is the main factor determining the structure of the ingot [1]. For low-alloyed, copper-bearing alloys with a high thermal conductivity, the grain size is determined by the ingot solidification rate—the higher the solidification rate, the finer the structure. However, copper alloys that contain high concentrations of other elements and, thus have, a low thermal conductivity are characterized by a critical solidification rate, above which abrupt coarsening of structure occurs [2].

Authors' affiliations: Kiev Polytechnical Institute, USSR.

A USh-114 installation with a liquid charge, which provides the most favorable conditions for dehydrogenation of the alloy [3], was used for remelting. A salt-based slag containing 85% cryolite and 15% fluorspar has a high refining ability for aluminum bronzes [4]. The same salt system was chosen as the basis for a number of slags used for ESR of copper and its alloys [5]. A drawback of this salt melt is the volatility of the aluminum fluoride ($AlF_3$), which is formed as the result of thermal dissociation of cryolite. The flux melting point is around 1,270 °K, which should provide the optimal degree of overheating of the slag layer.

Consumable electrodes are remelted in a water-cooled copper freezer mold (the diameter of the electrode was one-half that of the ingot). The alloy for the electrodes was made by radiofrequency induction (RFI) melting and poured into disposable sand–ceramic molds. Castings were subsequently machined to shape.

The effectiveness of the remelting process was determined on the basis of the gas and nonmetallic impurity content and the thermomechanical characteristics that most fully reveal the alloy's ME capability. These characteristics are the deformation limit, or the critical deformation that can be completely recovered $\epsilon_n{}^{100}$, and the reactive stress $\sigma_r$ that is, the stress generated in the alloy upon heating when it is mechanically constrained. Thermomechanical characteristics were measured in bending tests with four point loading.

The concentration of hydrogen and oxygen in the remelted alloy was determined by means of a "Leco" apparatus. The amount and dimensions of nonmetallic inclusions were determined by metallographic analysis using a "Kvantimet-720" apparatus and were based on the analysis of 500 sample areas on each polished cross section. Methodologies developed in the E.O. Paton Electric Welding Institute were used for these analyses of gas and nonmetallic inclusions. Alloy element concentrations and the martensite transformation temperature were also investigated. Macro- and microanalyses of specimens cut from ingots and electrodes were performed.

Numerous gas and shrinkage defects, located mainly around the axis of the electrode, were detected in longitudinal macrospecimens (Fig. 12.1). The microstructure is characterized by large, randomly oriented martensite plates (direct transformation of the alloy is complete at room temperature).

Ingots with a high-quality surface and a compact grain structure free from shrinkage defects (Fig. 12.2) were obtained using the following optimal remelting parameters; current 1.0 to 1.2 kA; voltage, 17 to 19 V; electrode feed rate, 0.7 to 0.9 mm/s. Upon withdrawal from the freezer mold, ingots were covered with a uniform thin slag skin that could be easily removed. The concentrations of alloying elements such as aluminum and manganese were identical in different regions of the ingot (head, center, tail, and periphery); that is, the difference in concentrations was lower than the resolution of the analysis method (0.05%).

Decreasing the remelting current led to an unstable remelting process and a corrugated surface; some of these corrugations contained cold laps. Increasing the current above the optimum value caused excess evaporation of flux and sharp variations (from 0.2% to 0.8%) in the aluminum content of the ingots obtained during different meltings but with electrodes of the same starting

V.K. Larin and D.F. Chernega

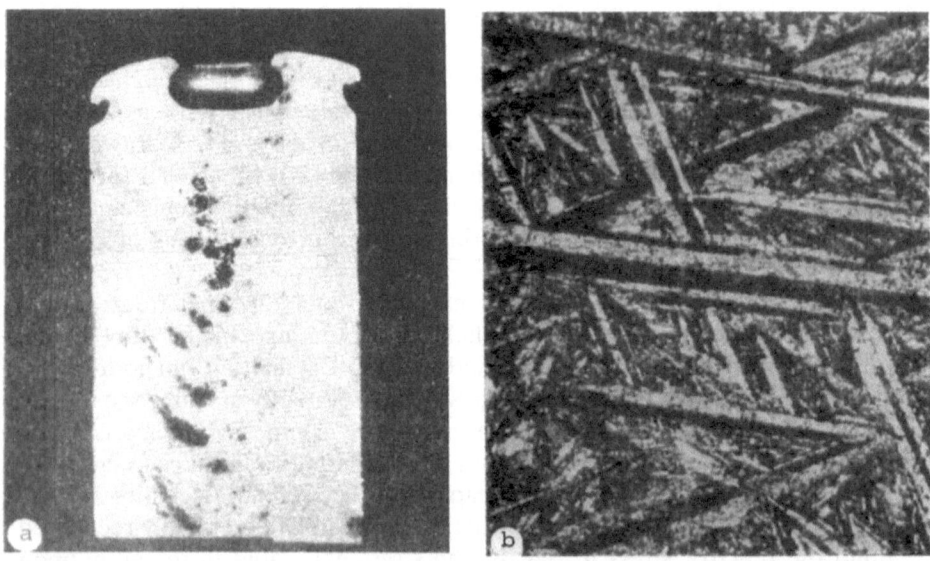

**Fig. 12.1.** (a) Templet and (b) microstructure of a bronze consumable electrode containing 11.9% Al and 5.0% Mn (× 450 magnification).

**Fig. 12.2.** (a) Templet and (b) microstructure of an electroslag-remelted bronze ingot containing 11.8% Al and 4.9% Mn (× 450 magnification).

**Table 12.1.** Characteristics of aluminum bronze with memory effect after electroslag remelting

| Alloying element (%) | | | NI*10: | Gas concentration | | Thermomechanical characteristics | |
|---|---|---|---|---|---|---|---|
| Al | Mn | Ak, K | (%) | $[O]^*10$,% | $[H]$, sm/100 g | $\epsilon_n^{100}$ (%) | $\sigma_r$ (MPa) |
| 11.90 | 4.95 | 327 | 3.05 | 0.82 | 4.30 | 2.25 | 455 |
| 11.80 | 4.85 | 340 | 0.22 | 0.52 | 3.55 | 3.65 | 525 |
| 12.60 | 4.05 | 320 | 2.32 | 0.75 | 4.65 | 1.80 | 385 |
| 12.50 | 3.95 | 338 | 0.23 | 0.47 | 3.55 | 2.75 | 465 |

Note: Electrode material, above line; ingot, under line.

chemical composition. This situation makes remelting of bronzes exhibiting ME impractical because it is almost impossible to provide the necessary thermal conditions for shape restoration.

With appropriate remelting parameters, a rather small, stable variation of alloying element concentration in bronze is obtained: The decrease in aluminum and manganese concentrations were 0.15% and 0.10%, respectively (Table 12.1). Martensite transformation temperatures for these alloys were higher than for the electrode alloy by 15 to 25 °K at most (Ak denotes the temperature at the end of the transformation in Table 12.1). This difference is transition temperature makes it possible to correct the composition of the charge used to prepare the electrodes, in order to make fine adjustments to the transition temperature of the remelted ingot alloy. However, the most effective method of adjustment is to test ESR bronze samples and then to correct the melt composition based on the resulting restoration temperature of the melt.

Remelting produces a more highly refined alloy than do other refining methods. The concentration of dissolved gases, especially oxygen, is lowered. The concentration by volume of nonmetallic inclusions is also reduced by almost

**Table 12.2.** Size distribution of nonmetallic inclusions in bronzes that exhibit memory effects that were refined by different methods

| Alloying elements (%) | | | | Number of | Number and size of NI in a 5-mm² area ($\mu$m) | | | | | | |
|---|---|---|---|---|---|---|---|---|---|---|---|
| Al | Mn | Refining method | NI (%) | NI (size $<1\ \mu$m) | 1.0–1.5 | 1.5–2.0 | 2.0–2.5 | 2.5–3.0 | 3.0–3.5 | 3.5–4.0 | Over 4.0 |
| 12.0 | 4.7 | — | 0.020 | 268 | 167 | 41 | 23 | 13 | 13 | 2 | 9 |
| 11.8 | 4.9 | MF | 0.008 | 110 | 53 | 22 | 17 | 6 | 5 | 3 | 4 |
| 11.9 | 4.9 | RF IMF | 0.006 | 73 | 26 | 12 | 20 | 3 | 4 | 3 | 5 |
| 12.5 | 4.0 | ESR | 0.002 | 19 | 10 | 4 | 5 | | | | |

Note: NI, nonmetallic inclusions; MF, melting under flux; RF IMF, RF induction melting under flux.

an order of magnitude, along with an improvement in their size distribution (Table 12.2). Nonmetallic inclusions larger than 2 $\mu$m were not observed in ESR alloy ingots, whereas in the starting material and in alloy refined by other methods, these accounted for one-fifth of all inclusions over 1 $\mu$m.

ESR changes the alloy structure to improve both the thermomechanical deformation ($\epsilon_n^{100}$) and the stress characteristics ($\sigma_r$) of bronzes. This is a property of ESR ME alloys only. For example, investment castings of the same bronze that was refined by using different salt melts (including a mixture of cryolite and fluorspar) could only recover from large deformation upon heating. The restoration of shape was accompanied by insignificant changes in reactive stress, which were probably related to a coarsening of the alloy structure—a characteristic of investment casting.

Thus, ESR can be successfully used to refine aluminum bronzes exhibiting ME and to improve their basic thermomechanical characteristics.

# References

1  B. Chalmers, "Solidification theory," in Metallurgiya, Moscow, p. 288 (1968)
2  O.D. Moldavski, T.A Stomahina, and M.I. Vainshtok, "Use of electroslag remelting for the production of copper based alloys," Non-Ferrous Metals 6, pp. 64–67 (1977)
3  Yu.V. Latash and B.I. Medovar, "Electroslag remelting," in Metallurgiya, Moscow, p. 238 (1970)
4  V.P. Zhuravlev, V.Ya. Schekaturov, and V.Ya. Filatov, "Refining of Br. AZhNMts bronze using molten flux," Casting Industry 4, pp. 16–17 (1973)
5  I. Yu. Lyuty and Yu. V. Latash, "Electroslag melting and refining of Alloys," Naukova dumka, Kiev, p. 187 (1982)

# 13 Thirty Years of Electroslag Remelting Steel Production at the "Dneprospetsstal" Plant

## I.G. Vodeniktov and Yu.G. Gabuev

The 30th anniversary of the first commercial electroslag (ES) ingot production at the "Dneprospetsstal" plant in Zaporozhye was celebrated in May 1988. Since the introduction of electroslag remelting (ESR) at the plant, continuous efforts have been made to increase processing capacity and to develop techniques for remelting a wider variety of steels. Further developments in ESR techniques and equipment were carried out jointly with research institutes, especially with the E.O. Paton Electric Welding Institute—the founder of ESR.

Until 1965 the plant was equipped with two three-phase furnaces—a DSS-1 unit and a P-951 unit built at the plant. The DSS-1 furnace was equipped with two units working in turn and was used to produce round ingots with a diameter up to 425 mm or square ingots with a cross section of $310 \times 310$ mm, and weighing up to 1.1 tons. The P-951 furnace was used to produce ingots $350 \times 350$ mm in cross section and weighing 1.65 tons.

For several years these furnaces have also been used for research on the technological capabilities of the process. This research revealed serious problems with power distribution in three-phase circuits. As a result, the furnaces were converted to single-phase operation, which improved their technological characteristics and yielded higher quality steel.

In 1966 a specialized ESR shop and new OKB-905 and OKB-1065 furnaces were put into operation. The new furnaces were designed to produce larger, square ingots, measuring $415 \times 415$ and $565 \times 565$ mm, and weighing up to 4 tons. The furnaces were used to investigate and to develop electrical parameters for remelting different types of slags under commercial production conditions. The problems studied included choosing steel types to be used for remelting; developing remelting methods to employ solenoids for slag and metal pool rotation; testing a bifilar power supply; and converting to process automation.

The accumulated experience in high-quality steel production and a skilled staff helped to solve the problem of obtaining high-quality ESR slabs. The first bifilar furnaces of the U-436 type for production of slabs weighing 13 tons were but into operation in 1967. Because of the expanding need for ESR steel sheet, two other new ESR furnaces (16VG-N1 and 20VG-M1), which were capable of producing ingots weighing up to 20 tons, were built at the plant. Their consumable electrodes consist of two slabs placed with their broad sides facing each other in the mold.

Authors' affiliation: A.N. Kuzmin "Dneprospetsstal" Electrometallurgical Plant, Zaporozhye, USSR.

The introduction of these new slab furnaces made it necessary to develop various auxiliary equipment. In order to use a process with a liquid charge, a crucible type ladle and equipment for siphoning the molten slag into the freezer mold and other devices were designed and built at the plant.

The tightening of requirements on ESR slab quality stimulated an intensive research effort, which led to the development of a bifilar circuit for remelting. This research also resulted in the building of new equipment and the introduction of techniques for remelting sheets and slabs in movable molds; in the development of a new method for paired assembly of electrodes (PAE) for remelting; and in the development of a multicomponent flux with increased basicity, along with remelting techniques using this flux. In addition, studies were performed on an automatic control system for ESR in a movable, T-shape freezer mold, in which inductive sensors were used to measure the height of the molten metal pool.

In recent years, research has aimed at improving furnace design, testing and analyzing the effect of various active media on the molten slag and metal, and automating the remelting process. At present, most furnaces are equipped with ARShMT-type automatic controllers that monitor and control the melting rate and the electrical resistance between the electrodes.

Investigations of and commercial tests on the modulation of the input power to the slag layer, when remelting slabs weighing up to 13 tons and using a two-electrode bifilar circuit, were performed. In addition, methods to measure and to control the melting rate by means of magneto-anisotropic weight sensors inserted into the furnace bottom plate were studied.

Production of ShH15SG and 40HN2MA steel in a freezer mold with a cross section of $565 \times 565$ mm, and a flux with increased basicity, was introduced in a furnace for remelting high-quality steel. This reduced the concentration of non-metallic inclusions by factors of two to seven. The use of a flux with increased bacisity for ESR of ShH15ShD, instead of an ANF-25 flux, made it possible to make the process more universal and to reduce the total number of fluxes used.

In order to increase the yield of the final product when electrodes obtained from an ingot weighing 3.6 tons were used, a $420 \times 530$-mm freezer mold was designed at the plant (the remelted ingot can weight up to 3 tons). Remelting of $240 \times 310$-mm electrodes in this freezer mold increased the yield of useful alloy by 700 kg over that obtained in a $415 \times 415$ mold.

The technology of additional alloying and injection of steel during ESR by means of the "Doza-3" injector constructed at the NIIPTmash was developed and further improved. ESR with reuse of up to 40% of the ANF-6 flux was introduced for the production of a number of constructional steels.

These are the main achievements in the further development of ESR at the A.N. Kuzmin "Dneprospetsstal" plant during the last 30 years.

Trends for further development in ESR at the plant are as follows: replacement and modification of existing ES furnaces; the use of cast electrodes instead of rolled stock; further improvement of the method by additional process modification of the slag and the molten metal, particularly by the use of injectors; and increased automation of the process.

# 14 Electroslag Remelting at the "Elektrostal" Plant

M.M. Klyuev, V.P. Kubikov, A.A. Pokrovski,
V.P. Stepanov, and K.Ya. Fedotkin

The "Elektrostal" plant currently uses several single-phase, OKB-905-type, electroslag remelting (ESR) installations that produce two sizes of round ingots, 0.25 and 0.42 m in diameter and 0.5 and 1.4 tons in weight respectively, and an OKB-906, three-phase installation that produces 4.2-ton ingots, 0.7 m in diameter.

The single-phase furnaces use a solid charge, whereas the three-phase furnaces are charged by siphon pouring of liquid slag that is initially melted in a crucible-ladle. The introduction of up-hill charging at the plant was a first in the world of metallurgy and was accomplished jointly with the E.O. Paton Electric Welding Institute. The technique has been patented, and licenses on up-hill charging have been sold in the West.

This ESR technique is still being perfected. For example, a method was recently developed for charging molten slag, which had initially been deoxidized in a crucible-ladle directly into a freezer mold or into a combination crucible-ladle mold. This led to additional improvements in the bottom end of the ingot (a reduction in nonmetallic inclusions, the degree of oxidation, etc.) and more uniform properties throughout the ingot.

Consumable electrodes, 0.19 m in diameter and 8 m in length, that were produced by the SCCM installation are used with various size freezer molds. Either three separate ingots with a diameter of 0.25 m or a single ingot with a diameter of 0.42 m can be produced from one electrode. To obtain ingots weighing more than 4 tons in the three-phase furnace, three consumable electrodes are simultaneously remelted into a freezer mold.

The process flow of steel production at the plant is as follows: electric arc (or induction) furnace–semicontinuous casting machines–remelting equipment (ESR, electric arc remelting). The introduction of this technology increased the final product yield by 10% to 15%, and the labor productivity by 20% to 30%, with increases in profit of 2 million rubles/year. This technology allows the production of specialized, hard-to-work carbon and alloy steels, including heat-resistant and ultra high-speed steels and steels susceptible to cracking.

A classification scheme was developed for carbon and alloy steels cast by semicontinuous casting machines and remelted using ESR, which permitted thorough analyses of each type of steel. The results of these analyses, after checking, were generalized for all steels of this type. Such an experimental approach was efficient and saved time and resources.

---

Authors' affiliation: The "Elektrostal" metallurgical plant, USSR.

**Table 14.1.** Concentration of nonmetallic inclusions in 13H11N2V2MF and 12H25N16G7AR steels

| Steel type | Steel description | Average content of nonmetallic inclusions (arbitrary units) | | | |
|---|---|---|---|---|---|
| | | Oxides | Sulfides | Brittle silicates | Spheroids |
| 13H11N2V2MF | Initial with SCC | 2.50 | 0.77 | 2.44 | 1.05 |
| | ESR | 0.85 | 0.68 | 0.84 | 1.01 |
| | Initial mold casting | 2.05 | 0.83 | 2.30 | 1.17 |
| | ESR | 1.02 | 0.64 | 0.81 | 0.68 |
| 12H25N16G7AP | Initial with SCC | 0.90 | 0.54 | 2.50 | 2.25 |
| | ESR | 0.65 | 0.50 | 0.63 | 0.60 |
| | Initial mold casting | 0.50 | 0.67 | 3.00 | — |
| | ESR | 1.06 | 0.57 | 0.74 | — |

Note: SCC, semicontinuous casting; 13H11N2V2MF and 12H25N16G7AR are designated EI961 and EI835 steels in State Standard 1787–70.

The joint efforts of the TsNIIChM, the E.O. Paton Institute, the VIAM, and the "Electrostal" plant led to the use of the combined continuous casting and ESR technique for over 200 high-alloy steel types at the "Electrostal" plant. Characteristics of the refining process in the course of ESR of continuously cast steel were studied in research carried out with the above institutes. It was determined that this technology greatly increases the refining effect. In spite of having larger oxide inclusions at a higher concentration than rolled electrodes, cast electrodes yielded steel with equivalent properties after remelting, as shown in Table 14.1.

Studies of the refining mechanism showed that the physical and chemical interactions of slag and alloy in the course of ESR are important. In addition, oxide inclusions were eliminated. In ESR of chromium and nickel–chromium stainless steels, it was found that there is an equilibrium between the aluminum concentration in the molten alloy and in the $Al_2O_3$ in the slag layer at the metal–slag interface. Based on these findings, a single-slag run was introduced for remelting the initial alloy, with a shortened refining period, since part of the refining process was transferred to the ESR operation. This increased the productivity of the initial refining step by 5% to 10%.

In the course of investigations carried out jointly with the TsNIITMASh, a modifying effect of ESR on the steel was discovered. This effect is caused by an interaction between the slag and the alloy components, especially with such chemically active components as titanium and aluminum. The modification effect depends on the chemical composition of the slag and the alloy and on the process temperature. The initial calcium concentration in H20N8T3 alloy is 0.013%, the final calcium concentration is 0.023% when a slag containing 5% CaO is used. It increases to 0.035% with a CaO concentration of 30%. Modification by various metals increases the ductility of the steel. As an example, modification has been used in the ESR of large ingots of 36NH alloy by adding magnesium to the initial

**Table 14.2.** Effect of magnesium injection on the final product yield of electroslag-remelted 36NH steel

| | Magnesium concentration (%) | | |
|---|---|---|---|
| Initial melting | ESR under a MgO-containing flux | Concentration in steel after ESR | Product yield (%) |
| — | — | 0.012 | 46 |
| 0.015 | — | 0.012–0.016 | 62 |
| 0.015 | 15 | 0.017–0.021 | 83 |

Note: In the initial melting, the amount of introduced magnesium was calculated only.

alloy and subsequently remelting this alloy under a manganese-containing flux, which significantly increases the ductility of the remelted alloy (Table 14.2).

Magnesium, rare earth metals, titanium, and other metals have been studied for use as modifiers during remelting. Modifiers are added as oxides and fluorides in combination with reducing agents and are introduced into the melt through injector feeders.

Investigations of the effects of reducing agents on the refining of steel have led to the development of consumable electrodes that do not need prior surface conditioning. When a liquid charge is used, the slag is reduced by aluminum, calcium, etc., in the crucible-ladle 10 to 15 min before it is poured into the mold.

An automatic injector is used to introduce deoxidizers during remelting. A proportioning device that was developed and built at the plant permits the introduction of predetermined quantities of various powder mixtures, including deoxidizers. The technique of introducing of reducing agents into slag significantly increases its refining ability and improves steel quality. In addition, the slag retains its refining properties and can be reused repeatedly (Table 14.3).

Studies on the adjustment of the steel's chemical composition in the course of ESR are of practical and scientific interest. In a three-phase furnace, chemical composition can be optimized by remelting three different, appropriately alloyed, consumable electrodes; in single-phase furnaces, the steel is alloyed during remelting.

Techniques for ESR of heat-resistant aluminum and titanium alloy steels that employ an equilibrium flux composition have been developed; they allow the

**Table 14.3.** Slag composition (%) after remelting of 12H12N2 steel

| Technique | $SiO_2$ | CaO | $A_2O_3$ | $FeO + Fe_2O_3$ |
|---|---|---|---|---|
| Without reduction of slag | 7.6 | 8.2 | 16.8 | 0.7 |
| With reduction of slag | 2.1 | 5.3 | 23.8 | 0.2 |
| Initial slag | 2.5 | 4.2 | 24.6 | 0.3 |

Note: The reminder of the slag is $CaF_2$.

**Table 14.4.** Change of sulfur concentration in electrolytic manganese after electroslag remelting

| | Sulfur concentration (%) | |
|---|---|---|
| Manganese characteristics | Before ESR | After ESR |
| MR-O electrolytic manganese | 0.062 | 0.009 |
| Manganese flakes | 0.120 | 0.012 |

reuse of flux, along with methods for additional alloying of these steels with aluminum and titanium injection during remelting.

A multistage ESR process that produces high-quality steel with reduced power consumption has been developed jointly with the E.O. Paton Institute and is also of great interest. At present, multistage ESR is regulated by an automatic control system based on KVPU 01.02 sequencers. In the future, a minicomputer will be used to control other equipment such as proportioning injector-feeders that measure and regulate the slag oxidation potential, regulate the slag level in the mold, regulate the elevation of the mold, and so on.

Another new technique of great practical importance is that of increasing the ingot length in stationary molds by elevating the mold.

Tests on a method of regulating ingot solidification by cooling the freezer mold with helium were done in collaboration with TsNIIChM. The test results showed a 12% to 15% increase in ESR productivity.

Studies on electroslag refining of manganese[1] were performed jointly with the E.O. Paton Institute. The process that was developed reduces the sulfur concentration in manganese by 80% to 90% (Table 14.4). The use of ESR manganese in the production of bimetallic thermostat components improves metal quality and decreases losses due to spoilage.

Work planned for the future includes developing new slag compositions; controlling the atmosphere in the mold; controlling the redox potential of the slag; and introducing alloying elements, reducing agents, and inoculants both before and during ESR. From an economic point of view, the most promising technology is to melt semifinished products and subsequently improve their properties by means of furnaceless processing and ESR. The combination of vacuum induction melting and ESR is very promising for the superalloys used in the aerospace industry.

To realize the above goals, it is necessary to develop reliable auxiliary equipment (injectors, protective devices, etc.), to be used at practically all electroslag installations. The design and construction of these devices are important. The new generation of ESR furnaces will have to have provisions for controlling the atmosphere in the working space of the freezer mold.

---

[1]Yu.V. Latash, V.A. Yakovenko, and I.Yu. Lyuty, "Electroslag refining of manganese," Chapter 11 of this volume.

# 15 New Developments in Electroslag Remelting at the Zlatoust Metallurgical Plant

A.B. Pokrovski, G.A. Hasin, V.I. Lazarev, L.A. Hrustalkov, V.A. Pozdnyakov, and B.M. Kukartsev

The introduction in 1961 of a three-phase furnace at the Zlatoust Metallurgical Plant brought about significant changes and the development of the existing Electroslag remelting (ESR) technology [1]. In particular, installations with three-phase transformers were replaced by single-phase transformer installations; a technique was worked out for remelting square ingots in molds with a square cross section, which made the ingots suitable for use in blooming mills; the solidification process of electroslag (ES) ingots was studied; a remelting technology was introduced that used consumable electrodes cast in semicontinuous casting machines; and laboratory methods for determinating the conductance, the viscosity, and melting points of slag melts were put into practice. A methodology for an approximate calculation of optimum melting conditions [2] allowed the reuse of up to 60% of the ANF-6 flux as a charge for the melting furnace. The chemical composition and physical properties of the used flux were adjusted to their original values by adding alumina and reducing agents. In addition, for one group of steel types, it was possible to adjust the optimal concentration of alumina by mixing ANF-1 and ANF-6 fluxes. These developments and measurements made it possible to improve steel quality and to make the remelting process more economical and efficient.

In recent years the main development trends in the field of ESR have been in the areas of improving steel quality (reducing in the course of remelting; inoculating/reinforcing the ingot) and increasing efficiency by lowering production costs for consumable electrodes, using fluxes with different compositions, and investigating new electroslag technologies (ESR with a rotating electrode, etc.). The Zlatous plant was also active in these areas.[1] The results of this research resulted in the adoption of a technology for the complex reduction of the slag in the course of remelting, and made it possible to obtain constructional steel ingots with a fine-grained structure (five to eight points on a relative scale) and also to adjust the concentration of certain alloy components. Inoculation of the alloy demonstrated the possibility of eliminating longitudinal cracks in ESR ingots with a square cross section fabricated of 30HGSA and other crack-sensitive steels. It has been recommended that this technique be put into production. Research work is ongoing on the use of inoculation during ESR to increase the homogeneity of the crystal structure of electroslag (ES) alloys.

---

[1] S.M. Krylov, Yu.V. Pogulyaev, L.V. Oyakonova, A.I. Maryunin, and B.A. Balashov participated in the research.

---

Authors' affiliation: Zlatoust Metallurgical Plant, USSR.

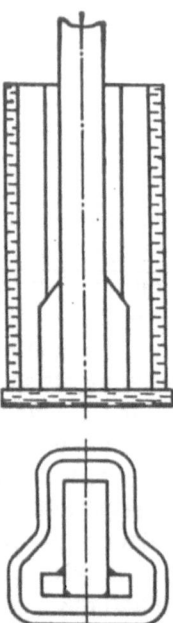

**Fig. 15.1.** Positioning of the consumable electrode in the mold to produce the upper block of a 3-ton forging hammer.

The fabrication of blocks for drop hammers weighing 3 and 7 tons was successful in both single-phase and three-phase (three electrodes remelted in a single mold) installations. Remelting is done in freezer molds with an internal cross-sectional shape that is identical to the shape of the hammer block (Fig. 15.1). Manufacturing hammer blocks by ESR eliminated the need for the purchase expensive semifinished products, reduced labor costs of machining, and increased the block life to the point of complete wear-out.

More efficient power consumption was obtained during ESR of high-chromium 20H13, EI961, EP56 steels, which contain up to 4% nickel because less ANF-6 flux was used. It was shown that it is possible to reduce power consumption by 40 to 190 KW-h/t while maintaining productivity. These steels were selected according to their thermal and physical characteristics and chemical composition, and remelting was performed in 390 × 390-mm molds.

A smaller decrease in power consumption was observed when concentrations of chromium (Fig. 15.2a) and austenitizing elements (Fig. 15.2b) (expressed as nickel equivalent, according to the Scheffler diagram) were increased. This ESR technique, in which a reduced quantity of flux is used, is now being used for mass production.

An integrated production technology for ESR of EP452 [3] and EP654 steels alloyed with titanium and aluminum that was developed at the plant results in stable, low losses of titanium a good surface finish. In addition, the titanium

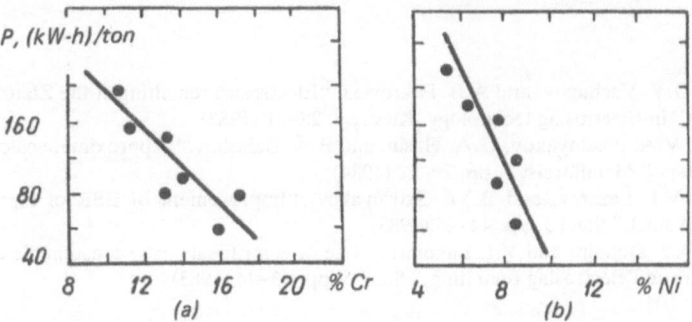

**Fig. 15.2.** Dependence of decreased ESR power consumption $P$ on the concentration of (a) chromium and (b) austenitizing elements in steel.

concentration along the ingot length was held constant due to the use of refining fluxes prepared at the plant.

Studies are being conducted to increase the yield of the final product by reducing the amount of cropped material from the head end of ingots produced in a process in which control over the termination of remelting is improved. This closer control is achieved through improved circuits for regulating the input current and voltage. Test results were favorable, and commercial production tests are planned. It was also proposed that the process of termination of remelting be automated. It must be admitted, however, that achieving a minimal slag depth at the end of ESR is hampered by a lack of voltage regulation in the range of 20 to 25 V in the power transformers of ES furnaces.

At present, it is possible, in principle, to use centrifugally cast electrodes in ES installations [4]. The necessary preparatory work is now being carried out so that a centrifugal casting installation can be built at the Zlatoust plant to fabricate consumable electrodes.

Studies of the quality of ES steel produced by ESR with a rotating consumable electrode are being continued. The advantages of this method, in comparison with conventional ESR, are its increased productivity, up to 30%, and its reduced power consumption, up to 20%.

The production of hollow ES ingots in installations with a movable mold and a bottom-located mandrel is in the planning stage. These installations will also be used to obtain shells that will subsequently be converted to bimetallic castings by vacuum arc remelting.

In recent years, 21 research projects were completed at the plant, independently and through joint work with the UkrNIIspetsstal, the VIAM, and the Metallurgical Department of the Chelyabinsk Polytechnical Institute in Zlatoust. Putting these new techniques into mass production has resulted in earnings of 128,000 rubles and material savings amounting to 161 tons of steel, 92 tons of ANF-6 flux, and 550,000 KW-h of energy. The quantity of recycled flux has been increased to 19%.

# References

1  G.A. Hasin, G.V. Vachugov, and A.B. Pokrovski, "Electroslag remelting at the Zlatoust metallurgical plant," in Electroslag technology, Kiev, pp. 28–31 (1983)
2  V.I. Lazarev, V.A. Pozdnyakov, G.A. Hasin, and B.A. Balashov, "Approximate calculation of
   ESR parameters," Metallurgist *3*, pp. 20–22 (1984)
3  G.A. Hasin, V.I. Lazarev, and B.Ya. Skornyakov, "Improvement of ESR of high-titanium
   H12N20T2-III steel." Steel *5*, pp. 44–45 (1983)
4  V.V. Loza, D.P. Dolinin, and V.I. Lakomski "Use of centrifugally cast consumable electrodes
   for vacuum arc and electroslag remelting," Steel *5*, pp. 43–44 (1983)

# 16 The Fabrication of Small Preforms for Machine Components by Permanent Mold Electroslag Casting in Factories of the Ukrainian Light Industry

K.K. Fishman, A.V. Zhalnin, B.Ya. Spivak,
N.I. Yasko, and Yu.V. Orlovski

Until a few years ago, it was thought that permanent mold electroslag casting (PMESC) could only be used for fabricating products weighing more than 5 to 6 kg. However, most components used in light industry are relatively small and weigh from 0.5 to 5 kg. Usually these castings are made by die or investment casting. Such castings are of rather low quality, due to casting defects, for example, internal and surface flaws, blowholes, or pores.

A technology for producing small castings (0.5 to 5 kg) from ferrous and non-ferrous metal scrap by means of PMESC was developed at the CSPKTBlegprom UkSSR and the E.O. Paton Electric Welding Institute.

The task of pouring small size castings is complicated by the fact that in PMESC the molten metal is poured into an open-topped mold. Thus, in order to obtain castings of a precisely determined height, accurate proportioning of the metal is necessary; this is not always possible in practice. The greatest difficulties are encountered when molten metal is poured into multisectional molds.

To eliminate such difficulties, a permanent mold, with so-called "pockets" for collecting superfluous molten metal, was designed (Fig. 16.1).

In the fabrication of small castings, proper positioning of the casting in the mold is of great importance to reduce runner losses. The shrinkage hollow in the casting (Fig. 16.2) and in the casing preform (Fig. 16.3) is relatively small.

To produce small components weighing up to 1 kg, an investment casting technology using ceramic molds was developed at the CSPKTBlegprom UkSSR. To increase the precision of the process, the slag is separated from the metal immediately before it is poured into the mold. The pouring device consists of a lined lid with an opening that is placed on the crucible; it permits complete separation of the slag from the alloy as it is poured into the mold.

The CSPTKBlegprom has developed a new method of fabricating disposable foundry molds from used flux, for use in both small-scale and mass production of castings for spare parts. After being melted, the flux is poured into a metal flask containing inserts of appropriate dimensions. When the slag has solidified, the inserts are withdrawn and the foundry mold is ready for use. The use of these disposable molds reduces the initial cost of various spare parts produced by PMESC.

During the XI to XII five-year plans, 12 specialized PMESC shops were set

---

Authors' affiliations: K.K. Fishman, A.H. Zhalnin, B.Ya. Spivak, and N.I. Yasko, CSPKTBlegproma UkSSR; Yu.V. Orlovski, E.O. Paton Electric Welding Institute, Academy of Sciences, UkSSR.

**Fig. 16.1.** Permanent mold with a proportioning "pocket."

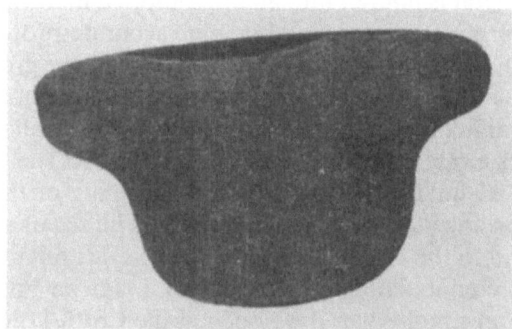

**Fig. 16.2.** Flange preform produced by PMESC in a multisectional mold (longitudinal cross section).

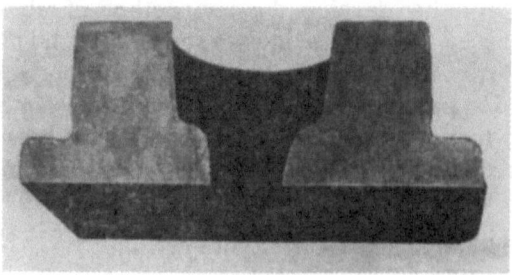

**Fig. 16.3.** Preform of a casing component produced by PMESC, with a ceramic core (longitudinal cross section).

up at various light industry manufacturing associations in the Ukraine. A PMESC technology to fabricate 87 types of small components was developed and put into mass production. For each kind of component, the most efficient PMESC technique was chosen. This involved taking into account the component shape, the duration of the production run, the required steel quality, and the grade of steel. The use of PMESC components in equipment manufactured for light industry increased the equipment life span 2 to 2.5 times, reduced the time required for equipment repair, reduced consumption of the starting rolled stock, and permitted more efficient use of the machinery. The net earnings due to the introduction of PMESC amount to from 800 to 1,200 rubles/ton of castings.

# 17 Sectional Strip Mill Rolls with Electroslag-Cast Outer Linings Having a Variable Chemical Composition Across the Width of the Roll

V.L. Matetski, V.A. Nikolaev, V.P. Poluhin, V.V. Chernyh, A.D. Chepurnoi, V.Ya. Saenko, and L.K. Leshinski

At present, research on development new methods and materials for the production of rolls used in rolling mills is carried out both in the Soviet Union and abroad. One of the promising production methods is electroslag casting (ESC) [1]. The advantages of rolls produced by ESC are reduction of material and labor costs per ton of product; the possibility of mechanizing and automating the process and creating materials with a guaranteed chemical composition; obtaining uniform wear of the working surface by means of a continuously variable chemical composition; increasing the wide range of materials that can be used in roll liner production in one installation (e.g., cast iron and low-alloy, low carbon and hypereutectoid steel); increasing the roll service life by means of optimal grain orientation during solidification; and increasing rolling mill productivity, due to prolonged roll service life.

In the Soviet Union, where ESC originated, several methods of roll production are used. The rolls are used mainly in the iron and steel industry. It has been determined that forged rolls can be almost entirely replaced by cast steel rolls, which are simpler to produce and have a longer service life. A problem of considerable importance concerns the production of rolls that vary in chemical composition across the width of the roll so that uneven wear during use is reduced.

So-called linear wear occurs unevenly across both back-up and working rolls. In a complement of rolls in a rolling mill, three stages of wear are observed: The first is a break-in stage, followed by a linear wear stage, and ending with catastrophic wear stage. The break-in stage is characterized by a nonlinear dependence of wear depth $\Delta D$ or $\Delta R$ on the number of cycles, after which the wear depth becomes linear until the beginning of the critical wear (catastrophic) stage. It is well known that the linear wear stage (which is longest) is characterized by good surface contact conditions, reduced heat generation, and the best flatness and minimal thickness variation of the rolled strips. A decrease in the duration of the break-in stage increases the duration of the linear wear stage and thus the length of service of a complement of working and back-up rolls. Since surface roughness, hardness, and rolling ability during this period are optimal, prior break-in increases roller service in a stand.

Authors' affiliations: V.L. Matetski, V.A. Nikolaev, and V.P. Poluhin, Moscow Institute of Carbon and Alloy Steels, USSR; V.V. Chernyh, Ministry of Heavy Industry, USSR; A.D. Chepurnoi, Manufacturing Association "Zhdanovtyazhmash," USSR V.Ya; Saenko, E.O. Paton Electric Welding Institute, Academy of Sciences, UkSSR; L.K. Leshinski, Zhdanov Metallurgical Institute, USSR.

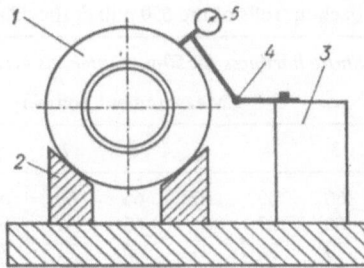

**Fig. 17.1.** Schematic diagram of the set-up for measuring the wear profile of rolls (see the text for identification of parts).

The accurate mating of surfaces is one of the main prerequisites for increasing the service of roll complements and improving the dimensional accuracy of rolled strips. Investigations of the wear of working and back-up rolls in various rolling mills, based on the wear relation $U = 1/\Delta R$ have led to the development of a semigraphic method of calculating the required hardness distribution across the width of the roll. In this relation, $U = $ constant is the condition for even wear.

The wear of back-up rolls of the second, third, and fourth stands of a type 650 strip mill was analyzed at the "Zaporozhstal" plant. The cross-sectional profile of rolls was determined after their removal for scheduled maintenance. Measurements were taken after the rolls had cooled to ambient temperature (Fig. 17.1).

To obtain profile measurements, the roll (1) is positioned in the supports of a surface grinder (2). A dial indicator (5), with a resolution of 0.015 mm, is fixed on a special rod (4) that is attached to the grinder carriage (3). The measuring rod is designed to allow a wide range of indicator (5) motion. Before measuring, the roll was positioned such that the runout measured on the thrust faces did not exceed 0.1 mm. Measurements were made at 50-mm intervals across the roll width.

The least amount of wear was observed at the ends of rolls, close to the thrust faces. The reasons for this nonuniform pattern are discussed in detail in the literature [2–4].

In addition to dimensional measurements, the hardness of the outer linings was also determined when they were removed for maintenance (Table 17.1). The Shore hardness of as-received rolls is 60 to 70. As is evident from the table, work hardening of forged rolls mainly takes place close to the ends, whereas the middle is relatively free of work hardening. Because of this, spalling occurs primarily at the roll ends.

As follows from the above discussion, the roll wears unevenly along its width. This results in poor mating between the working and back-up rolls, which leads to a redistribution of contact stresses.

On the basis of studies of back-up roll wear in a type 650 strip mill carried out at the "Zaporozhstal" plant and the application of a semigraphic analysis

**Table 17.1.** Surface hardness of back-up rolls of the 650 mill at the "Zaporozhstal" plant

| Initial hardness | Shore hardness (at 50-mm intervals across the width) | | | | | | | | | | | |
|---|---|---|---|---|---|---|---|---|---|---|---|---|
| | Measurement numbers | | | | | | | | | | | |
| | 1 | 2 | 3 | 4 | 5 | 6 | 7 | 8 | 9 | 10 | 11 | 12 |
| 65–68 | 75 | 67 | 66 | 69 | 67 | 66 | 68 | 66 | 66 | 67 | 71 | |
| 65–67 | 73 | 67 | 66 | 67 | 66 | 66 | 67 | 68 | 67 | 68 | 71 | |
| 68–70 | 72 | 69 | 70 | 72 | 71 | 66 | 71 | 70 | 68 | 68 | 71 | |
| 68–70 | 69 | 70 | 72 | 69 | 69 | 68 | 68 | 68 | 68 | 68 | 72 | |

method, specifications were developed for an electroslag remelting (ESR) process for the production of roll linings.

The development of a new manufacturing technique to produce rolls that wear evenly across their width became possible due to studies of the interdependence between wear resistance, concentration of alloying elements, and number of load cycles.

As shown in Fig. 17.2, for identical chemical composition and hardness of the working and back-up rolls, the varying frictional contact results in uneven wear of the back-up roll. The distribution of the necessary amount of additional alloying along the length of the ESC was calculated on the basis of the measured wear distribution. The results of the semigraphic calculation of the required chromium concentration were compared to results obtained by direct chemical analysis of the casting (Fig. 17.2). The roll liners were produced under commercial conditions, and the calculated results were used to control the addition of chromium and to take into account the variation in remelting parameters with

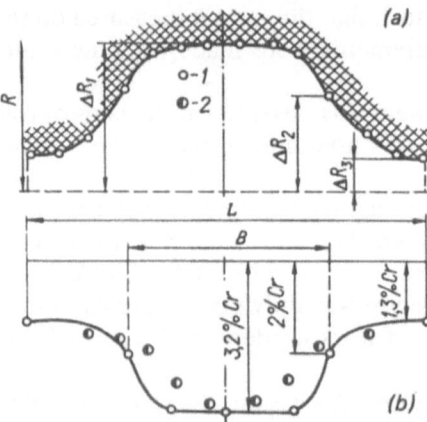

**Fig. 17.2.** Semigraphic method for the determination of the concentration profile of chromium alloying according to wear measurements (a) and chemical analysis (b) in order to provide uniform wear of cast rolls (L, roll length; B, rolled strip width).

variations in chromium concentration along the casting length (from 1.3% at the ends to 3.2% in the middle).

A trial batch of roll liners for the four-roll mill at the "Zaporozhstal" plant was produced at the MA "Zhdanovtyazhmash" in a USh-100 ESC installation. A total of eight rolls was fabricated in the first batch: three with a uniform chemical composition along the casting length and five with a varying composition.

The "Electrostal" plant assisted in developing preliminary and final heat treatments for the ESC roll liners. The preliminary heat treatment involved double normalizing and tempering and was performed before stripping. After the preliminary heat treatment, the Brinell hardness was no greater than 260, and after the final heat treatment, the Shore hardness was between 67 and 70. The castings were subsequently machined.

Analyses of ground rolls revealed that the lining material was dense and defect-free. Ultrasonic inspection was performed using three instruments: indirect, at 2.5 MHz; prismatic (55 degrees) at 2.5 MHz; and prismatic (40 degrees) at 1.8 MHz, with a depth sensitivity of 4 mm. No internal flaws were detected in this depth range. On the basis of studies of back-up rolls of the type 650 mill at the "Zaporozhstal" plant and test casting of linings for these rolls, a temporary technological specification 24-1-14-202-79 was developed for the production of ESC roll linings.

Service tests of this type of roll showed an increase of hardness of 1.5 times over sectional rolls with forged linings.

# References

1   B.E. Paton, B.I. Medovar, and L.M. Stupak "Electroslag technology," Naukova dumka, Kiev, p. 256 (1983)
2   P.I. Poluhin, Yu.D. Zheleznov, and V.P. Poluhin, "Rolling of steel strip and roll service life," in Metallurgiya, Moscow, p. 512 (1972)
3   M.M. Safyan "Rolling of wide strip," in Metallurgiya, Moscow, p. 160 (1989)
4   V.P. Poluhin, V.A. Nikolaev, and M.A. Tylkin, "Reliability and service life of rolls for cold rolling," in Metallurgiya, Moscow, p. 503 (1971)

# 18 The Introduction of Electroslag Technology at the Manufacturing Association "Novokramatorsk Manufacturing Plant"

E.A. Matsegora, A.T. Chepelev, G.N. Svistunov, A.S. Gavrishko,
V.Z. Kamalov, A.I. Borovko, Yu.A. Grushko, A.S. Volkov,
B.B. Fedorovski, Yu.G. Emelyanenko, C.Yu. Andrienko,
and V.Ya. Maidannik

Since 1959 the Novokramatorsk Machinery Plant has been collaborating with the E.O. Paton Electric Welding Institute on developing electroslag (ES) techniques for the production of critical parts. The involvement of skilled institute personnel and the substantial research capabilities of the institute allowed the rapid introduction of new technologies into commercial production. The ESR-2 furnace for producing electroslag-remelted (ESR) ingots weighing 2.5 tons and the ESR-10M furnace with a capability of up to 14 tons were put into operation at the plant in 1959 and in 1962, respectively. At the time, they were the largest such furnaces in the world. The three-phase ESR-10M unit can handle electrodes with a diameter from 330 to 350 mm and a length of up to 7 m. This type of electrode could be produced at the plant only by forging. That resulted in a low yield of ESR steel of 60% of the ingots used to produce electrodes.

Long delays in electrode production occurred because the forging press was in high demand for other products. Thus, even though the performance of the three-phase ESR-10M furnace was satisfactory, it was decided to convert it to a single-phase operation in order to produce large ingots more efficiently. The remelting of a single, 14-ton electrode made it possible to increase its diameter to 850 mm and to reduce its length to 3.5 m. These electrodes were produced primarily by casting into specialized molds.

Increasing the consumable electrode diameter to such an extent was an important, albert technically uncertain step. At the time, it was believed that increasing the electrode diameter to over 500 mm and the melting rate to over 500 kg/h would change the drop-by-drop flow of electrode material to a steady stream and that this would short out the electrode to the molten metal pool. This belief cast doubt on the possibility of ESR of large-diameter electrodes and of producing large ingots.

Experiments on remelting 850-m diameter electrodes in the ESR-10M furnace were carried out with the E.O. Paton Institute; they showed that the ESR of this

Authors' affiliations: E.A. Matsegora, A.T. Chepelev, G.N. Svistunov, A.S. Gavrishko, V.Z. Kamalov, A.I. Borovko, Yu.A. Grushko, and A.S. Volkov, MA "Novokramatorsk Machinery Plant," USSR; B.B. Fedorovski, Yu.G. Emelyanenko, C.Yu. Andrienko, and V.Ya. Maidannik, E.O. Paton Electric Welding Institute, Academy of Sciences, UkSSR.

type of electrode increases the flow of droplets, which, however, do not coalesce into a continuous stream, and that remelting has the same stability as in the ESR of small-diameter electrodes. These experiments showed that it would be possible to produce large ingots by ESR, and they became the basis for designing ESR furnaces with a capacity of over 60 tons.

The employment of a single-phase power supply and cast consumable electrodes sharply increased the efficiency of the process and the quality of the ESR steel.

Electroslag remelting (ESR) came to be appreciated at the plant for its simplicity, high productivity, and high-quality product.

In the course of 30 years, there has been a wealth of experience in the production of ESR steels for the manufacture of high-reliability components. Thus, the first ESR products, a gas turbine shaft and rotors of 3H13 and 13H11N2VMF steel, were appreciated by their users because the component's service life increased by factors of 8 to 10.

ESR steel was used in the production of the first Soviet 800-MW turbine for the Slavyanskaya Hydroelectric power plant. Its use completely eliminated machining operations on steam line parts, and increased their reliability. When these components had been manufactured from conventional electric furnace steel, 75% of them were rejected because of unsatifactory microstructure, contamination by nonmetallic inclusions, and poor mechanical characteristics.

The advantages of EST were particularly apparent in production forgings for large disks and rotors for use in power plants. The service conditions of these components demand that the steel contains no detrimental impurities, that it has low concentrations of evenly distributed nonmetallic inclusions, and that it has the required physical properties, especially high-temperature resistance. The latter requirement is especially significant for components that are subjected to dynamic loading. As a result of our investigations, ESR steel was recommended for the manufacture of high-power steam turbine disks.

ESR is a very promising technique in the manufacture of rolls used in rolling mills. Rolls are used under conditions of alternating loading, high contact stresses, and high friction. The main cause of roll failure is contact fatigue of the working surface. Rolls manufactured from ESR steel have a higher contact-fatigue strength at their working surface, due to their increased density and homogeneity and reduced concentration of nonmetallic inclusions. These characteristics prolong roll service life. The average service life of ESR steel rolls is 1.5 times longer than that of conventional steel rolls.

Electroslag technology (EST) holds great promise for scrap recycling. A practically scrap-free ESR technology, which uses worn-out rolls for consumable electrodes, was successfully tested and introduced at our manufacturing association. This technology yields yearly savings of 3,060 tons of steel and an equivalent fuel saving of 790 tons, which costs 538,000 rubles.

The high quality of ESR steel rolls, their isotropy across the barrel diameter, and the high contact-fatigue strength of their core all combine to allow repeated reuse of rolls after grinding and surface hardening.

For more than 20 years, EST has been used in the production of 9H2MF and

60H2CMF steel working rolls that are used in the cold rolling mills of the
"Zaporozhstal" plant, the Zhdanov metallurgical plant, and the Magnitogorsk
metallurgical plant.

Numerous problems concerning the production of high-speed steel were
solved in research work carried out at the Novokramatorsk manufacturing
plant. Consumable electrodes are produced by casting high-speed steel scrap.
The average strength of tools fabricated from ESR steel is 1.5 to 2 times higher
than that of tools fabricated from conventional steel.

In addition to high-speed steel, ESR tool steel production has been introduced
at the plant. The strength of cold forging dies of ESR H12M steel is two times
greater than that of conventionally produced dies. The yearly earnings from
ESR of high-speed steel are 67,000 rubles.

One significant step in the development of EST was the construction of an
automatic control system for ESR furnaces in the industry's largest ESR shop.
This system is designed to control and optimize remelting conditions of large
ingots. Since 1985 this control system measures and processes remelting parameters of the ESR-10G and 6ESR20SV furnaces, displays current parameter
values and instructions, and provides a printed melting report. In 1985, 800 ESR
ingots were produced under automatic control. Due to improvements in the
properties of nickel–chromium steels, 125 tons of alloy were saved, a saving of
80,000 rubles.

In 1986, for the first time in world metallurgy, the automatic control system
(ACS) of ESR-10G and 6ESR20SV furnaces functioned in the control mode
with complete computer control of remelting.[1] The ESR techniques developed
at our plant are equal to the best in the world, and in some cases, they are more
advanced.

In the current five-year plan, studies were started on the use of ESR to produce preforms having dimensions close to those of the final product (e.g., sectors
of coiling reels, transmission shafts, thin-walled cylinders, components for large,
heavily loaded gear drives). This will save 370,000 rubles and 510 tons of alloy,
and will also decrease labor input by 12,000 person-hours. To realize these
plans, the MA NKMZ and the E.O. Paton Institute have planned to establish a
specialized shop on the basis of an USh-126 installation with a serial electrode
feed.

The establishment of a shop to produce molten steel around a DSP-12 electric
furnace will permit the development of such promising methods as casting with
ES metal, which would reduce the amount of trimmings from 30% to 8% and
yield a steel with quality close to that produced by ESR. This would save 600
tons of alloy a year. In addition, this method would increase the weight of single
castings 1.5 to 2 times and reduce their initial cost by 100 rubles/ton.

---

[1]V.I. Mahnenko, E.D. Gladk, and Yu.A. Skosnyagin, "Computer Automation of ESR
Furnaces," pp. 38–44 of this book.

# 19 Electroslag Casting at the Manufacturing Association "Cheboksar Commercial Tractor Plant"

L.Z. Tsygurov, A.G. Galkov, E.M. Ofitsersov,
V.N. Kuznetsov, Yu.M. Mironov, V.G. Kovalev,
N.I. Atamanyuk, and Yu.Yu. Petelin

Some components produced from forgings for the manufacture of heavy tractors at the MA "Cheboksar Commercial Tractor Plant" (CCTP) are characterized by a low utilization factor of metal (UFM) (0.17 to 0.35), which results in the inefficient utilization of scarce constructional steels (40H, 33HS, 38HS, 20HN3A, 40GMFR, etc.) and increases machining costs.

The "cone" method developed at the E.O. Paton Electric Welding Institute produce preforms for driving wheels of UT-330 tractor [1] was the first application of electroslag technology (EST) at the MA CCTP. Later improvements in the design of the mold bottom plate and in melting conditions made it possible to decrease the casting height from 220 to 165 mm, which reduced machining allowances. The service life of the copper freezer molds exceeds 4,000 melting runs.

In 1978 a pilot electroslag casting (ESC) shop equipped with an R-951 furnace, an EODTsN 4800/10 transformer, and a U-560 flux melting furnace was put on-line. With the introduction of the pilot shop, a part of the need for ESCs for heavy tractor components was satisfied.

In 1983 the first ESC production line started operation, with a capacity of 5,000 tons of castings/year. It is equipped with ESR-2.5L and ESR-0.25VG furnaces and U-560 flux melting furnaces. Two requirements for ESC are specific to the MA CCTP: First, there is a wide variety in the type of components (gears, hubs, axles), and their weight (from tens to hundreds of kilograms), which are produced by ESC; second, these parts are produced in quantities of up to several thousand a year.

A new subdivision was organized at the MA CCTP, where, together with the Electrical Engineering Department of the Chuvash State University (and PKII "Promtractor," 1977 to 1980), research is carried out on new ESC techniques.

The wide variety of component configurations necessitates the use of several different ESTs: direct ESC and ESC with fusion of the poured steel into bar stock or forgings that are attached to the side walls or bottom plate of the freezer mold [2]. Requirements for high-volume production place additional restrictions on process productivity, casting quality, and equipment reliability.

Besides being used to produce drive wheels, conventional ESC is used for

Authors' affiliations: L.Z. Tsygurov, A.G. Galkov, E.M. Ofitserov, and V.N. Kuznetsov, Manufacturing Association "Cheboksar Commercial Tractor Plant," USSR; Yu.M. Mironov, V.G. Kovalev, N.I. Atamanyuk, and Yu.Yu. Petelin, I.N. Ulyanov Cheboksar State University, USSR.

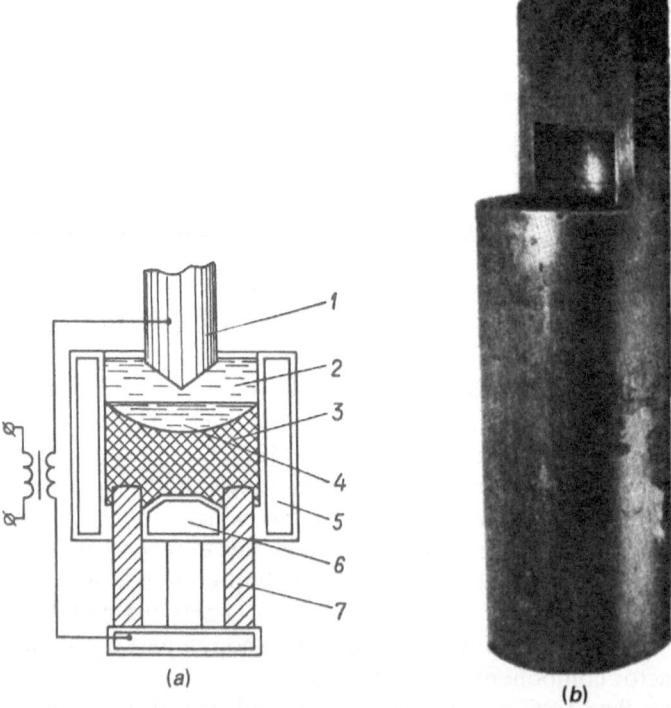

**Fig. 19.1.** (a) Schematic diagram of ESC in which remelted steel is fused onto the end of a tubular preform: (1) Consumable electrode; (2) slag layer; (3) remelted steel; (4) molten metal pool; (5) freezer mold; (6) mold bottom plate; and (7) tubular base. (b) Photograph of a typical component produced by this technique.

pinion gears in which the socket cavity is formed by a thin sheet steel, disposable mandrel.

ESC with fusion onto inserts makes it possible to produce more complicated shapes and reduces both machining allowances and melting time. At present, three techniques are in use: ESC with initial fusion between the remelted metal and a preform that is placed in the mold; a process in which the remelted metal encircles a cylindrical preform that is concentric with the freezer mold; and ESC in which the remelted metal is fused to preforms placed in openings in the side wall of the freezer mold.

The technique of ESC with initial fusion was tested using preforms in the form of shafts and pipes placed in the bottom plate of the mold (Fig. 19.1a). This method is used for producing tubular components (Fig. 19.7b).

The second ESC technique is used to produce castings for hubs. The remelted metal forms the flange and encircles the tubular base (Fig. 19.2). Since the base is preheated by the slag before it comes into contact with the molten metal, there is good fusion between the flange and the cylinder, and hot cracks do not occur during shrinkage of the cast alloy. Proper selection of the slag bath dimen-

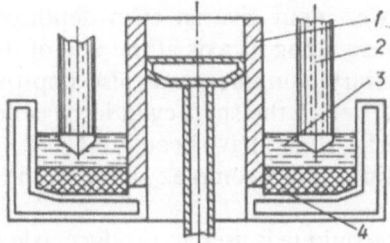

**Fig. 19.2.** Schematic diagram of ESC wherein a flange of remelted steel is fused to the exterior of a tubular preform for fabricating hubs. (1) tubular preform; (2) consumable electrode; (3) slag layer; and (4) remelted alloy.

(a)                                    (b)

**Fig. 19.3.** Hubs fabricated by the ESC method diagrammed in Fig. 19.2. (a) Hub with a simple flange and (b) hub with a pinion blank.

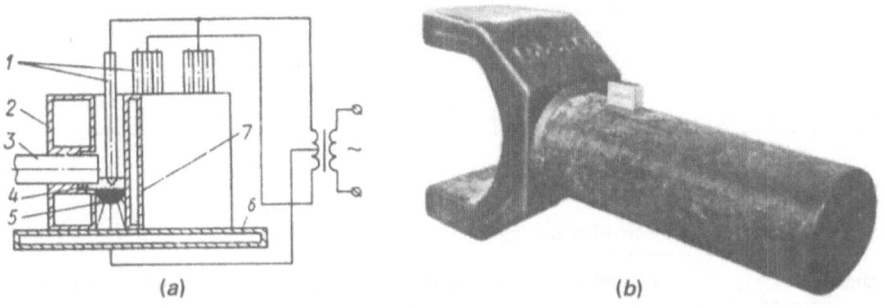

(a)                                    (b)

**Fig. 19.4.** (a) Schematic diagram of ESC with fusion of the remelted steel to the face of a preform that is inserted through the side wall of the mold: (1) consumable electrodes; (2) support panel of the freezer mold; (3) shaft preform; (4) slag layer; (5) remelted steel; (6) mold bottom plate; and (7) U-shaped mold.

sions and the electric power input give an even depth of the fused zone and minimize melting of the base along its axis at the end of the process to no more than 3 to 5 mm. The auxiliary equipment was also improved. A sprinkler-type cooling system for the interior of the shaft cylinder was developed. Over 3,500 hub castings of this type (Fig. 19.3a) have been produced by this ESC technique. It is also possible to obtain a more complex shape in the rim of the flange, as shown in (Fig. 19.3b).

The third type of ESC technique is used to produce axle castings. Remelting is done in a U-shaped freezer mold, and the remelted steel is simultaneously fused to the face of a horizontally situated shaft (Fig. 19.4a) in order to obtain the required component shape. The application of these ESC techniques provides a savings of 1.6 tons of alloy in the production of one tractor (Table 19.1).

The quality of parts obtained by means of EST from steels 40H, 20H, 33HS, 38HS, 20HN3A, which are used in tractor manufacture, was analyzed, and estimates were made of the variation in the chemical composition of the steel re-

**Table 19.1.** Savings of steel due to the use of electroslag casting in the production of large tractor parts

|  |  | Steel savings (kg) | |
| --- | --- | --- | --- |
| Part type | Part weight (kg) | Per component | Per complete tractor |
| Drive wheel | 248 | 550 | 1,010 |
| Axle | 104 | 177 | 354 |
| Sun gear | 52 | 17 | 34 |
| Hub | 60 | 97 | 194 |

**Table 19.2.** Reduction of manganese and silicon concentration in steel after electroslag casting

|  | Manganese (%) | | Silicon (%) | |
| --- | --- | --- | --- | --- |
| Steel grade | Concentration | Loss | Concentration | Loss |
| 40H | 0.59 | 6.8 | 0.28 | 21 |
| 20H | 0.61 | 13 | 0.27 | 20 |
| 38HS | 0.73 | 5.5 | 1.26 | 9.5 |

**Table 19.3.** Properties of electroslag castings of 38HS steel

| Specimen | $\sigma_v$ (MPa) | $\sigma_{0.2}$ (MPa) | $\delta$ (%) | $\psi$ (%) | KCU (MJ/m$^2$) |
| --- | --- | --- | --- | --- | --- |
| Bar stock |  |  |  |  |  |
| (diameter, 200mm) | 783 | 670 | 17 | 44 | 0.25–0.35 |
| Alloy from fusion zone | 700–745 | 700 | 12–16 | 30–31 | 0.45–0.35 |
| Remelted alloy | 755–782 | 551–592 | 28.7–28.8 | 48.2 | 0.49–0.52 |

melted under ANF-6 flux in order to determine the loss of alloying elements and to correct the composition of the consumable electrode (Table 19.2).

The strength of the remelted steel, and of the steel in the fusion zone, is similar to that of wrought, electric-arc steel. The properties of components of 38HS steel obtained by ESC with fusion are given in Table 19.3. These results are similar to those given in the literature [3].

The experience of the first ESR shop in the world's tractor manufacturing industry in producing a variety of components for heavy tractors shows that ESC is economically competitive with technologies in which forged products are used.

An analysis of the operation of existing electroslag (ES) furnaces shows that there is great potential for increasing the efficiency of the ESC process through greater mechanization and automation and the application of such new technologies as permanent mold electroslag casting (PMEC) and centrifugal electroslag casting (CESC). A widespread application of CESC at the MA is planned for the 12th five-year plan. At present, dismountable freezer molds are required to produce complex castings, which involves a long preparatory period of ESC. Thus, the coefficient of equipment usage is no greater than 0.3.

The standard output of a single worker is 64 tons of castings/year, which is more than the output in conventional conveyor foundries. The power consumption per ton of product is 1,200 KW-h, which is not greater than the power consumption for conventional casting.

In order to increase the efficiency of EST at the MA further, a number of problems remain to be solved, such as improving techniques, mechanizing and automating the main and the auxiliary operations, and coordinating the operation of the equipment and control systems.

## References

1  I.I. Kumysh, V.T. Desyatov, and Yu.B. Petrov, "Electroslag casting of drive wheels for commercial Tractors," in Problems of Electroslag Technology, Kiev, pp. 154–158 (1978)
2  B.E. Paton, B.I. Medovar, and G.A. Boyko, "Electroslag casting," in Casting metallurgy, NIImash, Moscow, p. 74 (1974)
3  B.E. Paton and B.I. Medovar, "Electroslag metal," Naukova dumka, Kiev, p. 680 (1982)

# 20 Applications of the Electroslag Process for Producing Cutting Tools from Scrap Tool Steel at Factories of the Ministry of Machine Tools of the USSR

V.A. Antonov, A.G. Miroshnichenko, L.S. Tkachuk,
E.V. Odegov, A.V. Zherebetski, V.B. Linetski,
N.A. Seroshtanenko, and G.A. Boyko

In recent years great attention has been paid to the effective use of resources in the Soviet Union. The factories of the Ministry of Machine tools of the USSR (MMT USSR) use high tonnages of scarce alloy and tool steels, which also produce large amounts of scrap (the amount of high-speed scrap is about 20,000 tons/year). The cost of tool steel rolled products varies from 500 to 8,000 rubles/ ton, but the scrap is bought by Vtorchermet USSR at prices ranging from 20 to 2,000 rubles/ton, depending on steel grade and type. Thus, developing low-scrap or scrap-free technologies for machine tool production is of great economic importance.

The technology of electroslag casting (ESC) that was developed at the E.O. Paton Electric Welding Institute produces higher-quality castings. Experiments on obtaining ESCs from tool production scrap, off-grade blanks, and rolling spoilage were performed in the ESC laboratory of the MRA NIISL. It was discovered that steel that was completely unsuitable for tool manufacture (because of defects such as inhomogeneity of properties, blow holes and shrinkage pipes, discontinuities, and porosity) were quite satisfactory after ESC. New ESC processes based on these results were developed for obtaining semifinished products for thread-cutting tools of H12 and H6VF steel and cold forging dies of R6M5 steel. For the commercial realization of these processes, ESC shops were built at the MRA NIISL, the Kamyshinsk machine tool factory, and the Maikop "Stanko-normal" plant.

At present, there is an insufficient variety of alloy steel semifinished products produced at the manufacturing plants of MMT USSR because the supply of rolled products does not fully satisfy the shape requirements of the finished products. Thus, the utilization factor of metal (UFM) is usually no greater than 0.4 to 0.5.

A wide range of tools is produced from ESR steel; these include broaches, forging dies, grooving rolls, threading tools, dies for pressure die casting, and cutting tools. Most of the castings weigh no more than 200 kg. The ESC shops are equipped with ESR-0.25VGI1 and ESR-0.25VGL1 furnaces, with a capacity

Authors' affiliations: V.A. Antonov, A.G. Miroshnichenko, L.S. Tkachuk, E.V. Odegov, A.V. Zherebetski, V.B. Linetski, and N.A. Seroshtanenko, Manufacturing Research Association NIISL, USSR; G.A. Boyko, E.O. Paton Electric Welding Institute, Academy of Sciences, UkSSR.

of 250 kg, which are built by the Ministry of Electrotechnical Industry. The associated auxiliary equipment was developed at the MRA NIISL.

Tool steel is remelted in the ESR-0.25VGI1 furnace, which is equipped with a 250-mm diameter aluminum freezer mold. Besides being used for experimental work, the ESC shop of the MRA NIISL produces batches of alloy steel ESCs to supply factories of the Odessa region (the A. Ivanov "Stankonormal" plant, the MA "Odessapochvomash," the MRA "Kislorodmash," the MPS Mechanical plant, the others). In addition, the shop is used to train personnel in the use of ESC equipment.

The advantage of the present shop and of those under construction lies in their capabilities in all types of electroslag technology (EST): ESC, electroslag welding (ESW), etc.

The EST shop of the Kamyshinski machine tool factory is shown in Fig. 20.1.[1] The shop equipment consists of an ESR-0.25VGL1 furnace, a vacuum-molding installation E-75A1 (developed at the MRA NIISL), and auxiliary equipment. The working area of the shop is 170 m², and its capacity is 100 tons of electroslag (ES) product/year. The main shop products are flanges with a cross section of

---

[1] The following engineers participated in the start-up work at the EST shop: S.V. Orlov, I.I. Avanesov, S.A. Efimenko, and E.I. Dyakov.

**Fig. 20.1.** The EST shop at the Kamyshinsk machine tool factory: (1) E-75A1 installation and (2) ESR-0.25VGL1 furnace.

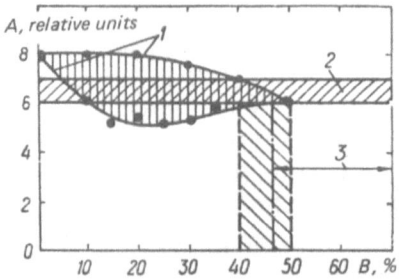

**Fig. 20.2.** Affect of the zircon concentrate content in the flux (B) on the carbide homogeneity (A) of tools fabricated from R6M5 steel. (1) ESC; (2) electric arc billet; and (3) instability zone of the ESR process.

$300 \times 230$ mm and a thickness of 30 mm and cast ES broaches of R6M5 steel. The service life of ESR broaches is 2.5 times that of broaches forged from ordinary electric arc steel, which is due to the reduced concentration of nonmetallic inclusions and the finer grain structure, with a uniformly distributed carbide network. These benefits are realized through optimized remelting conditions and by the use of a flux containing a zircon concentrate.

As shown in Fig. 20.2, the use of 40% to 50% zircon concentrate yields ES R6M5 steel having a carbide structure equivalent to that of R645 rolled stock. Metallographic analysis and machinability tests showed that the carbide is more highly dispersed and that the steel is denser and more machinable than steel produced by conventional ESR techniques.

Machining is the most labor-consuming process in the manufacture of equipment. In order to reduce labor input and increase UFM, a combined vacuum–electroslag process (CVESP) was developed. This process improves precision-cast ES products so that they satisfy the First Class precision requirements of State Standard 2009-55 and have a surface finish that falls in Classes 11–12, according to State Standard 25346-82.

The creation of the EST shop and the introduction of ES processing at the Kamyshinsk machine tool factory provides net earnings of about 147,000 rubles.

The Maikop "Stankonormal" plant produces standard machine parts of manufacturing.[2] Threads on fasteners are formed on thread-rolling machines by means of special rollers, dies, and segments.

The plant's ESC shop has one ESR-0.25VGI1 furnace. A combination of aluminum, copper, and steel freezer molds is used. A new technique for reusing scrap H12M and H12MF steel was developed for the manufacture of threading tools. Thread-rolling tools with a diameter of 50 mm were produced by two techniques: by ESC, in a 65-mm diameter aluminum mold; and by ESC, in the same aluminum mold, followed by forging.

---

[2]M.V. Belau and G.M. Chmyhov participated in the initial EST work at the Maikop "Stankonormal"

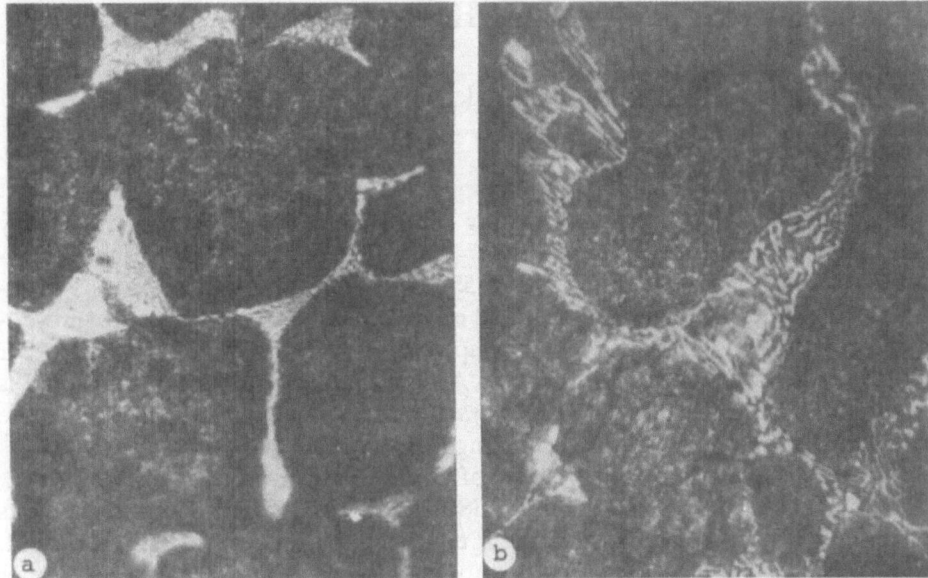

**Fig. 20.3.** Microstructure of specimens from (a) electroslag casting, and (b) electroslag casting that has been forged ( × 500 magnification).

It was determined that after the third reforging operation the strength of thread-rolling dies is equivalent to that of dies produced from conventional rolled stock. The strength of ES rollers is 2.5 to 3.0 times higher than that of conventional rolled stock. The microstructure of the ES cast steel corresponds to grade 5 of State Standard 5950-73; it consists of sorbite with a Vickers hardness of 318 to 345 and a torn network of lederburite eutectic that is distributed along the grain boundaries (Fig. 20.3a). The microstructure of the forged alloy consists of troosto-martensite (Vickers hardness 909) with a lederburite eutectic (Fig. 20.3b). After forging, the lederburite network remains, with a slightly coalesced carbide phase; here, coalescence of carbide inclusions is caused by a long preforging heating period. Apparently, overheating the steel in an oxidizing environment results in the formation of oxide precipitates along the grain boundaries, which significantly worsens the service characteristics of thread-rolling dies.

The earnings from the introduction of EST at the Maikop "Stankonormal" plant amounted to 137,000 rubles. Recycling of tool steel scrap and the production of tools with improved service at three manufacturing plants of MMT USSR yielded earnings of 400,000 rubles.

# 21 Electroslag Technology at the Manufacturing Association "Kolomensk Heavy Machinery Plant"

Zh.I. Yuzhanin and R.S. Dubinski

The development of electroslag casting (ESC) at the MA "Kolomensk Heavy Machinery Plant (KHMP) began in 1974. The first two stages of this work consisted of producing castings in type A-550 ESR furnaces; analyses of the properties of cast electroslag (ES) steel and comparisons with forged steel; and production of various castings weighing up to 3 tons in the welded steel freezer molds of the R-951UM ESR furnace. A detailed description of that period is given in the book *Problems of Electroslag Technology,* which is devoted to the 20th anniversary of electroslag remelting (ESR) (1978).

The third stage in the development of ESC of large preforms started in 1981, with the introduction of the USh-108 ESC installation. This installation consists of three ES furnaces, each of which is equipped with a separate freezer mold. The installation design also provides for melting steel into a common freezer mold using either two or all three furnaces. The maximum casting weight, which theoretically can be produced when all three furnaces are functioning, is 100 tons. The USh-108 installation is the largest ESC complex in the Soviet Union. Initially, because of small production quantities simple types of castings, such as plates and sleeves with a height of 1.5 m, were remelted in the steel freezer molds of the Ush-108 installation when it was first used commercially. Disposable, water-cooled steel cores were used for forming openings. To provide maximum versatility, adjustable molds consisting of welded steel components were developed.

The successful use of steel freezer molds for castings weighing up to 12 tons in a four-year period (1981 to 1985) proved the concept. However, there were also some problems connected with the use of these molds. Thus, it is necessary to give a more detailed description of the use of welded steel molds for ESC. The main drawback of these molds, compared to copper molds, is the increased likelihood of burn-through of the working wall and explosion of the mold.

Ten years' experience in the use of the stationary steel freezer molds in producing various castings weighing up to 3 tons in the R-951 installation demonstrates their workability.

To avoid significant thermal deformation of the molds, the thickness of the interior walls is no greater than 5 mm, and internal vertical ribs are used for stiffening. The distance between the ribs is not greater than 40 wall thicknesses. There are headers for uniform water distribution at the cooling water inlet and outlet.

Authors' affiliations: E.O. Paton Electric Welding Institute, Academy of Sciences; UkSSR.

In order to guarantee the safe operation of welded steel molds, well-known rules must be followed. In the plant it was found that steel freezer molds are capable of lasting up to 100 melting runs, and they offer certain advantages for ESR of single castings.

However, the use of steel molds in mass production significantly reduced the productivity of the installation because of time-consuming assembly and disassembly operations. In connection with this, from 1984 to 1985, a product assortment of similar ES castings was studied. At present, the casting assortment consists of cubical forging die preforms, weighing from 2 to 6 tons, and 7-ton roller preforms for continuous casting machines (CCM). These preforms are melted in sectional movable copper molds, in which direct current is used for the rollers and a bifilar power supply is used for the forging dies.

The forging die preforms are melted in adjustable molds that can produce castings with cross sections from $300 \times 300$ to $600 \times 1000$ mm. The consumable electrodes are made from $220 \times 220$-mm 5HNM tool steel castings produced by semicontinuous casting.

Experience in the field with cast ES forging dies at factories of the Ministry of Agricultural Machines Industry (the Minsk tractor plant, the Lozovsk forging-mechanical plant, etc.) showed that, after appropriate heat treatment, the service life of ESC dies is 1.5 to 2 times longer than that of forged ones.

The use of worn-out dies as raw material for consumable electrodes significantly improves the service life (two to three times, according to data from the Kolomensk diesel locomotive factory) of cast ES dies, compared to forged dies.

Through the use of a bifilar electric circuit and a freezer mold space factor of 0.4 to 0.5, it is possible to extend the life of the copper, water-cooled elements of the mold for up to 200 melting runs. The mold is repaired by electric arc welding.

The production of rollers, in which a direct circuit power supply and a mold space factor of 0.6 to 0.7 is used, is characterized by erosion of the mold top of up to 10 mm, after which repairs to build up the mold are required. In order to reduce wear, a water-cooled, steel extension is inserted into the mold. The slag surface is maintained within the extension by adjusting the remelting parameters. After 50 melting runs, neither the extension nor the mold showed visible signs of wear.

In the USh-108 installation, ordinary freezer molds with straight walls are used. Control of the slag level is performed visually and by means of sensors inserted into the mold from above. For the first time, a large, tubular, semifinished product with an external diameter of 1.455 m, a wall thickness of 70 mm, and a height of 2.5 m was produced in this installation. The weight of the O7H16N6 steel casting was 6 tons (Fig. 21.1).

Melting of tubular, semifinished products F760/F640 of DI-52 steel in movable adjustable molds is now being put into production.

Welded steel molds are widely used in the R-951UM installation at the ESC laboratory to produce small quantities of castings weighing less than 3 tons.

Production of tail spindles in a movable freezer mold has been initiated (Fig. 21.2). This 550-kg ESC has replaced a 900-kg forging ESC and has significantly reduced the amount of machining necessary.

**Fig. 21.1.** A 6-ton sleeve with an 1.455-m outer diameter and an 1.315-m inner diameter is 2.5 m long. It was produced in a movable mold from O7H16N6 steel.

**Fig. 21.2.** Semifinished tail spindles are produced in a movable freezer mold (casting weight, 550 kg; forging weight, 900 kg; steel, 20H).

Work is also proceeding on the development of permanent mold electroslag casting (PMESC) at the ESC laboratory. For that purpose, a special installation with a ladle-tipping device has been designed and built at the laboratory. The installation can remelt consumable electrodes weighing up to 1 ton. The ladle tipper is thyristor controlled in order to pour the ES metal into the mold at an optimal rate.

The use of ES, semifinished products has solved numerous problems of production of large components and has replaced the forgings previously supplied by other plants.

# 22 Electroslag Equipment at the "Sibelektroterm" Manufacturing Association

V.G. Zavyalov and A.G. Pomeshikov

The MA "Sibelectroterm" has been manufacturing electroslag (ES) furnaces for more than 20 years, starting in 1966. The first ES furnace produced at the association, the model OKB-1111, is still in operation at another manufacturing plant. Since that time, several dozen other furnaces built by "Sibelectroterm" have been installed in the Soviet Union and abroad.

As electroslag technology (EST) developed, new ES furnaces were built for various applications: furnaces for the production of forged and sheet ingots: installations for electroslag welding (ESW) of castings; furnaces for producing hollow tubular castings; and furnaces for proportional casting (Table 22.1). Most furnaces were developed jointly with the E.O. Paton Electric Welding Institute and the main institute of the thermoelectric industry subdivision of VNIITO.

The latest achievements in the field of electroslag remelting (ESR) were taken into account in developing new furnaces. For example, the use of a bifilar circuit increased $\cos\varphi$, compared to power applied to the electrode and bottom plate, and reduced the specific power consumption.

The use of a liquid charge technique, in which remelting is initiated by pouring molten flux into the mold, both speeds up remelting and improves the quality of the bottom end of the ingot. The MA also produces flux-melting furnaces with a crucible capacity ranging from 250 kg to 18 tons for preliminary flux melting (Table 22.2).

The development of remelting in movable freezer molds increased the potential for controlling the cooling rate of the exposed part of the ingot.

Service testing of mass-produced furnaces with the capacity to melt ingots weighing 10 and 20 tons was performed at "Sibelectroterm," which made possible improvements in the furnace design in parallel with their installation at end-user plants. Thus, the newer version of the ESR-10G-12 furnace was equipped with a 3,200-KV-A transformer instead of the original 2,500-KV-A transformer, as well as a more advanced ARShMT-2 control system. New features were added to the orginal furnace design, such as an injector-feeder, an electromagnetic stirrer for mixing the molten steel and slag, and a device for introducing inert gas into the melting zone.

Experience with previous designs was put to use in the construction of the ESR-10VG-11 furnace, which was exported to Czechoslovakia in 1985. This furnace is equipped with a low-frequency power source (0.5 to 10 Hz) to reduce inductive losses and to increase furnace efficiency. The furnace is also supplied

---

Authors' affiliation: MA "Sibelektroterm," USSR.

**Table 22.1.** Some characteristics of electroslag remelting furnaces built at the MA "Sibelektroterm"

| Furnace type | Maximum casting (weight, tons) | Type of electrode circuit | Transformer capacity (KV-A); number of transformers/ furnace (in parentheses) | Year of production |
|---|---|---|---|---|
| OKB-1111 | 60 | M,G | 7,500(1) | 1966 |
| U-436/6ShSP-4.12.9 | 9 | B,G | 3,500(1) | 1970 |
| OKB-1155A | 4.5 | M,G | 2,500(1) | 1972–1974 |
| ESR-10G | 8 | M,B,G | 2,500(1) | 1973–1979 |
| U-658/63ShS-2,5 | — | B,S | 3,600(1) | 1973 |
| ESR-16V1N1 | 16 | M,B,V,G | 2,500(2) | 1974 |
| OKB-1155B | 4.3 | M,V | 2,500(1) | 1975 |
| ESR-16VGI2 | 16 | M,B,V,G,P | 3,500(1) | 1975 |
| ESR-20VGI1 | 20 | B,V,G | 2,500(1) | 1975–1976 |
| ESR-150/6ShSTs-27-150 | 200 | B,V | 5,000(3) | 1977–1978 |
| EShO-200I1 | 200 | Three-phase pouring of molten metal into slag bath | 10,500(1) | 1977 |
| 6 ESR-20SV | 20 | M,B,V,G | 5,000(1) | 1978 |
| ESR-5LI1 | 5 | B,V,G | 5,000(1) | 1979–1981 |
| ESR-20VGI2 | 20 | M,B,V,G | 5,000(2) | 1979–1980 |
| ESR-5VGI1 | 5 | M,B,V,G | 2,500(1) | 1981–1984 |
| ESR-10VGI1 | 10 | M,B,V,G,P | 2,390(1) | 1985 |

Note: M, Monofilar circuit (electrode, bottom plate); B, bifilar circuit; V, movable freezer mold capability; G, stationary mold; P, capability of melting hollow castings; S, electroslag welding.

**Table 22.2.** Some characteristics of flux melting furnaces produced at "Sibelektroterm"

| Furnace type | Transformer capacity (KV-A) | Type of electrode connection | Maximum weight of melted flux (tons) | Year of production |
|---|---|---|---|---|
| U-360/6RK0-1, 0FS | 1,000 | M | 1.0 | 1970–1981 |
| U-560/6RK0-0, 7FS | 760 | M | 0.25 | 1963–1985 |
| OKB-1449 | 2,500 | B | 1.0 | 1974–1980 |
| U-360/6RK0-1, 0FSA | 2,500 | B | 2.5 | 1978–1980 |
| 6RK3-2, 5FSA | 2,500 | Three-phase | 18 | 1978–1979 |
| RK3-2FSI2 | 2,000 | | 5 | |

Note: M, monofilar circuit with graphitized electrode; B, bifilar circuit.

with both movable and stationary freezer molds. The design incorporates bifilar and monofilar circuits. The process is computer controlled by ShTsD-9702 devices, based on the V7 microprocessor.

The experience of the Swedish firm "SKF-steel' with the ESR-16VG-12 furnace helped in the development of the new 6ESR-20SV furnace. This furnace can produce solid and hollow round ingots in stationary or movable molds using either monofilar or bifilar power supply circuits. Remelting of one, two, four, or more electrodes is possible. The furnace can be modified to produce square and rectangular ingots, as well as ingots with more complicated cross sections. The movable molds and their bottom plates are constructed of bimetallic plates that are fabricated by explosive welding, which was found to be the best way to apply the clading steel reinforcing layer on the copper plates. The strength of threaded and welded joints in the steel layer is much greater than the strength of joints in the copper layer.

The ESR-20VG-12 furnaces were service tested to determine the optimal values of a number of parameters. In addition, the operating electric characteristics of the furnace were determined, including $\cos\varphi$, power loss, and the available useful power, and these values were used for design corrections. Strain measurements revealed the most highly stressed furnace elements and allowed the determination of safety margins in order to improve the service reliability of the furnace.

The results of service tests on working furnaces are very helpful, both in improving existing equipment and in developing new designs with optimized operating parameters.

In addition, the development of new furnaces necessitated the introduction of new manufacturing techniques at "Sibelektroterm," most importantly in the area of freezer mold production. Panel molds are fabricated from copper or bronze plates with a thickness of 80 to 90 mm. A new technique was worked out for drilling deep holes with diameters of 18 to 40 mm using specially designed drills to form water channels with lengths of up to 3 m. A method for machining the internal surfaces of the mold using fly cutters with specialized attachments was introduced. The preparatory-welding production was also considerably modified. A new technique for plasma cutting of the stainless sheet used in the manufacture of freezer mold jackets, electrode holders, movable carriages, and other ESR furnace equipment was perfected. Also, advanced methods of welding in inert gas and carbon dioxide atmospheres were adopted. With assistance from the VNIIESO and the E.O. Paton Institute, a plasma welding method was developed for assembling bronze and copper molds, for welding sheet, for making longitudinal and circular transverse welds, and for welding flanges. At the same time, plasma generators and welding techniques were perfected, increasing the quality of welded joints. The furnaces were put through simulated production tests before delivery, in order to make final adjustments and to ensure reliable operation.

This practice was later extended to other equipment built at "Sibelektroterm," such as arc furnaces and laser processing equipment.

In order to reduce the length of the research and development period, joint engineering groups were organized, incorporating personnel from "Sibelektro-term," the E.O. Paton Institute, and VNIIETO.

The plant's wide experience in the development and operation of ESR furnaces gives it the capability of building equipment that is competitive internationally.

# References

1  L.A. Volohonski, A.A. Nikulin, and A.G. Pomeshikov, "Commercial equipment for electroslag remelting and casting," Abstracts of the VIII All-Union Conference on Electrothermics and Electrothermal Equipment, Cheboksary, July 3–5 (1985)
2  Yu.A. Naryshkin, V.I. Belski, and A.G. Pomeshikov, "Equipment for Electroslag Technology," Electrotechnical Industry. Electrothermics 5, (1984)

# 23 The Use of Electroslag Technology in the Production of Electric Furnaces

L.A. Volohonski, M.A. Kisselman, A.A. Nikulin, and E.G. Protokovets

The method of electroslag remelting (ESR), which originated at the E.O. Paton Electric Welding Institute, is now the leading technology in commercial specialized electrometallurgy.

Electroslag (ES) furnaces, that produce high-quality castings of various sizes and shapes, which weigh several kilograms to 200 tons [1] are currently widely used in the Soviet Union and abroad.

Electroslag-remelted steel is used to make a wide variety of critical components, from miniature bearings to large, many ton rotors for high-power electric generators and turbines. The use of electroslag-melted steel is widespread due to its improved service characteristics, the result of fewer detrimental impurities and nonmentallic inclusions and an improved crystal structure (Table 23.1). In addition to further improvements of the process and equipment capacity, the continued development of ESR in recent years has been characterized by the introduction of new electroslag technology (EST) methods and the use of ES processes to produce heat-resistant and nonferrous alloys. Comparisons of the properties of ESR copper and conventionally produced copper are shown in Table 23.2.

Another widely used ES method is electroslag casting (ESC). This technique produces high-service castings and a shape and size close to those of the final product, which reduces the amount of machining and scrap losses. It is now used to manufacture a wide variety of products, including pipes, high-pressure vessels, valve casings, crankshafts, rolls, and large reinforcing rings.

The development by VNIIETO of specialized furnaces to manufacture com-

Table 23.1 The effect of electroslag remelting on the mechanical characteristics of 20HGS2N steel [2]

| Steel production technology | Specimen orientation | $\sigma_v$ (MPa) | $\delta$ (%) | $\psi$ (%) | KCU (MJ/m$^2$) |
|---|---|---|---|---|---|
| Open hearth | Longitudinal | 1,525–1595 | 11.0–11.6 | 50.0–54.8 | 0.88–0.97 |
|  | Transverse | 1,500–1,570 | 10.2–10.8 | 43.2–46.5 | 0.37–0.56 |
| ESR | Longitudinal | 1,560–1,580 | 12.0–12.2 | 50.1–51.0 | 1.03–1.21 |
|  | Transverse | 1,555–1,556 | 15.4–15.6 | 55.6–62.1 | 0.97–1.04 |

Authors' affiliations: L.A. Volohonski, M.A. Kisselman, and A.A. Nikulin, VNIIETO, USSR; E.G. Protokovets, MA "Azerelektroterm," USSR.

**Table 23.2.** The effect of electroslag remelting on the mechanical characteristics of M1 copper [2]

| Production technology | $\sigma_v$ (MPa) | $\sigma_s$ (MPa) | $\delta$ (%) | KCV (MJ/m$^2$) | Nonmetallic inclusions |
|---|---|---|---|---|---|
| Permanent mold casting | 170 | 70 | 18.0 | 0.6 | 40 |
| ESR | 174 | 88 | 46.0 | 1.47 | 42–44 |

plex ES products, which weigh from 250 kg to 5 tons (Table 23.3), has also helped move ESC technology into mass production.

The production of bimetallic preforms of plain and complex shapes [3–5] is made possible by another new EST method, ESRB and ESCB based on remelting a consumable electrode under a slag layer in an uncooled freezer mold. A newly developed method for obtaining cylindrical, tubular, and rectangular bimetallic semifinished products in either stationary or movable molds is ready to be moved into mass production.

The new method has been used to produced the following bimetal combinations: constructional steel/copper, bronze or brass of various compositions, and stainless steel/copper. When ESC techniques are employed, it is possible to obtain steel/bronze castings of a predetermined composition. This can be achieved by alloying the copper with the necessary components during remelting. Steel/copper castings have been produced with a diameter of up to 570 mm and a height of 2 m. An inside diameter of hollow castings as large as 150 mm has been produced by means of a "piercing" technology; here, the minimum thickness of the built-up copper layer is 5 mm.

In an interesting application of this method, bimetallic slabs are subsequently mechanically worked. At present, the maximum dimensions of slabs for rolling are 60 × 500 × 1000 mm (the steel is 15 mm thick). In addition, these bimetallic copper/steel products find applications in the manufacture of electric contacts and heat exchangers.

Analyses have shown firm bonding between the two metals, with failure under load occurring in the lower-strength metal. The service characteristics of components made from bimetallic products were studied in laboratory and commercial tests. The results demonstrated the high reliability and increased lifetime of these components.

**Table 23.3.** Characteristics of commercially produced electroslag casting furnaces.

| Furnace design | Maximum casting (weight, tons) | Transformer capacity (KV-A) | Furnace dimensions (mm) |
|---|---|---|---|
| ESR-0.25VGL | 0.25 | 630 | 5,400 × 2,800 × 3,720 |
| ESR-1.25L | 1.25 | 1,000 | 5,500 × 6,000 × 8,000 |
| ESR-2.5L | 2.5 | 2,500 | 6,600 × 6,000 × 9,150 |
| ESR-5L | 5 | 5,000 | 5,900 × 8,200 × 11,000 |

**Table 23.4.** Components used in electroslag remelting furnaces and production methods

| Part | Main requirements of the part | Production method |
|---|---|---|
| Drive screw for carriage positioning | High strength | ESR |
| Drive nut for carriage | High strength; low friction coefficient | ESR, ESC, ESCB |
| Rack | High strength; low friction coefficient | ESR, ESC |
| Gear | High strength; low friction coefficient | ESR, ESC, ESCB |
| Sprocket | High strength; low friction coefficient | ESR, ESC |
| Current carrying parts of electrode holder | High electrical conductivity; high strength | ESCB, ESC |
| Freezer molds and bottom plates | High thermal and electrical conductivity; high-temperature strength | ESR, ESRB, ESCB |

Analysis of the service characteristics of ES alloys and of the characteristics of various ESR methods have showen that one of the most promising applications of EST is in the manufacture of critical electric furnace components that will be used under conditions of high mechanical, electrical, and thermal stress and components of ESR and ESC furnaces, in particular (Table 23.4).

Comparisons of estimated production costs of a movable cylindrical freezer mold for the ESR-0.25VGL furnace are shown in Table 23.5 for three manufacturing techniques: the conventional method, ESC of a shell with a stationary core (mandrel), and electroslag casting of a bimetallic shell. As is evident from Table 23.5, the reduction of copper and stainless steel consumption, due to reduced machining losses, lowers the cost of bimetallic molds by 30% compared to molds produced by the conventional technique and by 10% compared to molds fabricated from ESC panels or a tubular ESC. A significant advantage of ESC, and of ESR of bimetallic products, is that it is possible to use small-diameter,

**Table 23.5.** Comparision of production costs of movable molds produced by competing techniques (%)

| Method | Material cost | Production cost | Total cost | Relative savings |
|---|---|---|---|---|
| Conventional | 25 | 75 | 100 | 0 |
| ESC | 15 | 75 | 90 | 10 |
| ESCB | 15 | 55 | 70 | 30 |

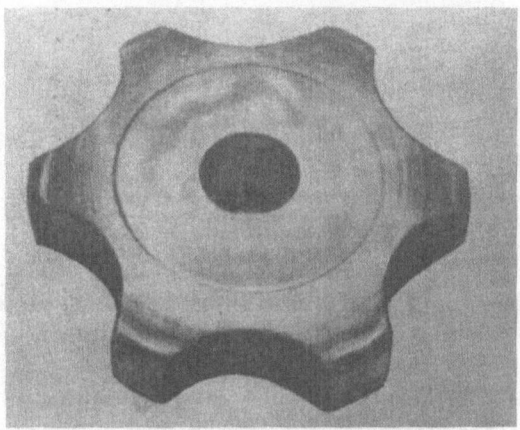

**Fig. 23.1.** Cast sprocket after partial machining.

consumable electrodes or even electrode scrap to produce large, complex components.

Various components used in ESR have been made at the VNIIETO by ESR. Some examples are the sprocket produced in a specialized sectional mold (Fig. 23.1); a thick-walled copper casting of a mold shell produced by remelting

**Fig. 23.2.** Copper shell for a stationary freezer mold produced by ESR.

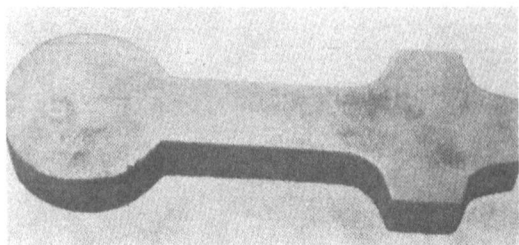

**Fig. 23.3.** Electroslag casting of a copper, current-carrying insert for an electrode clamp used in a remelting installation.

relatively thin electrodes (Fig. 23.2); and a copper, current-carrying insert for electrode clamps (Fig. 23.3).

The ESR of bimetals finds an especially promising application in the production of complex, labor-consuming freezer molds that are used in ESC installations. In this case, the steel part of the component stiffens the whole construction and also facilitates the welding and fastening of the sections.

The use of ESCs for electrothermal equipment is not confined to ESR installations. EST technology can be also used in the manufacture of components for various equipment that operates under high mechanical, thermal, and electrical stresses.

ESRB and ESCB technologies were used to produce bimetallic contact clamps for arc and ore-smelting furnaces; nipple units of combined water-cooled electrodes of PEShO installations, bottom electrodes of plasma-arc furnaces; freezer molds for various types of remelting installations; etc. [4].

# References

1  B.E. Paton and B.I. Medovar, "Electroslag furnaces," Naukova dumka, Kiev, p. 416 (1976)
2  B.E. Paton and B.I. Medovar, "Electroslag metal," Naukova dumka, Kiev, p. 680 (1981)
3  A.L. Andreev, G.S. Andreeva, and L.A. Volohonski, "Electroslag casting of bimetallic steel/copper products," Electrotechnical Industry. Electrothermics 6, 214, pp. 15–17 (1980)
4  A.L. Andreev, V.A. Bobin, and L.A. Volohonski, "Use of bimetallic products in the manufacture of parts for remelting installations," Electrotechnical Industry. Electrothermics 3, 231, pp. 7–8 (1982)
5  L.G. Puzrin and A.Sh. Gorodetski, "Electroslag building-up using a stationary freezer mold," Nukova dumka, Kiev, pp. 9–14 (1980)

# 24 Research at the Georgian Polytechnical Institute in the Field of Electroslag Technology

S.B. Yakoboshvili, I.Yu. Mogilner, D.M. Haradze,
G.Sh. Kobalava, G.G. Bikoev, and E.N. Hundadze

Investigations in the field of electroslag technology (EST) at the Georgian Polytechnical Institute (GPI) were started in 1980 at the Department of Metallurgy and Welding Techniques and at the laboratory of the student association "Specialized Electrometallurgy and Welding." At present, research and development work in this field is carried on at the Republic Welding and Specialized Metallurgy Scientific Center (RWSMS) of the State Committee on Science and Technology of the Georgian Soviet Socialist Republic (GSSR), which was established in the Welding Department of the institute.

The RWSMS laboratory has an EST shop with a modified A-550U installation, a tabletop USh-114 installation, and a mechanized flux melting installation that was built at the GPI.

The modified A-550U installation is equipped with a sliding rail current feed to the electrode (Fig. 24.1), which replaces the original 5-m cable. The set-up consists of a 10-mm-thick, current-carrying, copper busbar, which is fastened to a column, and a copper slider, which encircles the busbar and carries six copper brushes that provide reliable contact. Four years' research experience in using the A-550U installation with a sliding current feed shows its high reliability and ease of service. Power is supplied to the A-550U and USh-114 installations by a TShS-3000-1 transformer.

In the last few years, several projects on electroslag casting (ESC) of components for the manufacturing and metallurgical industries were undertaken at the laboratory. A detailed description of the equipment and the ESC technique of producing castings of lathe tool holders is given in the literature [1]. Techniques for casting tool holders with close-to-finish dimensions are being perfected in the ESC shop of the S.M. Kirov Tbilisi Machine Tool Plant. In employing this technique, the utilization factor of metal (UFM) is increased to 0.65, as compared to the 0.45 that was achieved with the method was used earlier at the plant.

An ESC shop is being started at the Rustavi Metallurgical Plant (RMP) [2] with the assistance of the institute's Welding Department. The shop has an A-550U installation and a TShS-3000-1 transformer. The new shop has developed a method for rebuilding worn-out rollers of the driving chain for a cooler in a sheet rolling mill. In this method [3], the worn-out, usually asymmetrically positioned hole in the roller body is filled (Fig. 24.2) by melting a consumable electrode prepared from scrap rolled stock; that is, in effect, ESR in a consumable freezer

---

Authors' affiliation: The V.I. Lenin Georgian Polytechnical Institute, Georgia, USSR.

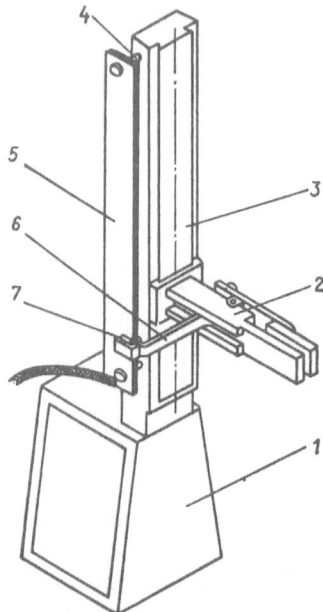

**Fig. 24.1.** Schematic diagram of the modified A-550U installation: (1) installation body; (2) electrode clamp; (3) column; (4) insulating standoff; (5) copper busbar; (6) slider; and (7) brushes.

mold is realized. After machining, the roller is ready for service. Rollers repaired by EST have been reused up to four times, which saves material and labor at the plant.

The technique for casting tube drawing mandrels that was developed by the GPI together with the E.O. Paton Electric Welding Institute was adopted in the ESC shop of the RMP. The patented arrangement consists of a water-cooled

**Fig. 24.2.** Rollers repaired by ESR.

**Fig. 24.3.** Electroslag-cast mandrels.

mold with a vertical parting line; castings are produced by remelting 12HZA steel electrodes with diameters of from 36 to 40 mm. Mandrel castings are close in shape to the finished product and require a minimum amount of machining (Fig. 24.3). The high-density remelted steel has a high structural homogeneity. The mechanical properties, chemical composition, and strength of the castings are comparable to those of drop forged mandrels. The superiority of the new technology, compared to the technology used at the plant previously (flame cutting, two-step heating and drop forging, scalping, final machining) is evident. In addition to its economic benefits, this fabrication technique has removed bottlenecks in tube drawing operations at the plant.

Joint research by the GPI and the RMP is planned for further EST developments and, in particular, for increasing the number of parts produced by ESC (various types of mandrels, tube-piercing points, etc.). It is also planned to move into production techniques for ESC of components having an axial variation in their properties or chemical composition. For this purpose, zonal alloying techniques are attractive, especially for the production of small- and medium-sized castings. This approach can produce composite castings with more advanced properties and characteristics, compared to castings produced by conventional means.

In order to realize overall control over the solidfication of ESCs, intensive research is being done at the Welding Department of the GPI and the RWSMS Center. The main topics of research are concerned with new ESR methods, that use specially shaped electrodes; the remelting of consumable electrodes that consist of short pieces fo unconnected rod; and studies aimed at improving the structure of the casting bottom when a process with a solid slag charge is used. A mathematical model of electroslag remelting with control of solidification is being developed.

# References

1   S.B. Yakoboshvili and I.Yu. Mogilner, "Electroslag technology," Collection of papers devoted
    to the 25th anniversary of electroslag remelting, edited by B.E. Paton, Kiev, pp. 135–137 (1983)
2   T.K. Giorgobiani and I.Yu. Mogilner, "Electroslag technology shop at the Rustavi Metallurgical
    Plant," Georgian Polytechnical Institute, pp. 44–48 (1984)
3   S.B. Yakobashvili and I.Yu. Mogilner, "Repair of sheet-rolling mill rollers by means of electro-
    slag fusion," Problems of Specialized Electrometallurgy *1*, pp. 21–24 (1986)

# 25 Developments in Electroslag Technology at the Manufacturing Research Association "NIIPTmash"

V.V. Chernyh, I.K. Marchenko, I.E. Blohin, G.A. Molodan,
V.I. Oleinichenko, P.P. Rudometkin, G.D. Lyubchenko,
V.V. Bezhin, and A.M. Litvinov

The efficiency of electroslag remelting (ESR) and the quality of the remelted steel depend to a large extent on the methods used to obtain the consumable electrodes [1–3]. The MRA has constructed a semicontinuous casting machine to produce consumable electrodes that are either circular in cross section, with a diameter of up to 1 m; square, with dimensions of up to 800 mm a side; or rectangular, with a maximum weight of 45 tons, and a final product yield as high as 96% (Fig. 25.1).

In order to improve ESR alloy quality, the MRA developed a special, multipurpose, continuous injector that introduces predetermined amounts of inoculants, ferroalloys, flux, and other additives during remelting. Up to 12 different components, with a maximum grain size of 10 mm, can be handled at the same time. [4]. This injector was put into operation at the Novokramatorsk Manufacturing Association for ESR of rolls for the 1680 rolling mill of the "Zaporozhstal" plant. Worn-out rolls were used for consumable electrodes. The wear resistance of the ESR rolls was increased by a factor of 2.1, due to the use of inoculants and to the extra alloying of the 60H2SMF steel with ferrosilicon and ferrochrome. The same type of injector was introduced at the Bryanski Manufacturing Plant in order to improve the physical and mechanical properties of ESR steel by the addition of ferrovanadium and multicomponent inoculants. The introduction of these types of injectors at the above factories yielded savings of over 350,000 rubles.

A universal ESR-1 installation with a capacity of 1 ton was designed by the MRA "NIIPTmash." The distinguishing feature of this installation is an automatic electrode positioning device. The MRA is also studying the application of permanent mold electroslag casting (PMESC) to the production of parts for the heavy machine industry.

Techniques and equipment for the production of ESCs of rolls for multiple-roll mills, forging dies, diesel engine shafts and hubs, and coiler barrel sections were developed at the MRA to meet the needs of the MA NKMZ and the Gorki "Dvigatel Revolutsii" plants. The MRA also supplies the PO BMZ, the Kaluzhski "Transmash" plant, and "NIIPTmash" pilot plant with ESC equipment used in the production of railroad car axle shafts, rock crusher parts, various support plates, casings, diesel engine pins, bucket dredges, cast-forged rolls,

Authors' affiliations: V.V. Chernyh, Ministry of Heavy Machinery of the USSR; I.K. Marchenko, I.E. Blohin, G.A. Molodan, V.I. Oleinichenko, P.P. Rudometkin, G.D. Lyubchenko, V.V. Bezhin, and A.M. Litvinov, Manufacturing Research Association "NIIPTmash," USSR.

**Fig. 25.1.** Consumable electrodes with cross-sectional dimensions of 480 × 480 and 460 × 350 mm, and a diameter of 600 mm.

**Table 25.1.** Specific wear of working rolls (tons of product/mm of barrel wear)

| Roll barrel diameter (mm) | Roll material production method | Grade of rolled steel | USSR | | Germany | Sweden |
| | | | NIIPT mash | SKMZ | | |
|---|---|---|---|---|---|---|
| 48 | 110H6SV2MF, ESC | 79NM | 80 | — | — | — |
| | 110H6SV2MF, ESC | E330A | 26 | — | — | — |
| | 9H, forged | E330A | — | 3.6 | — | — |
| | 160H5V7F5K5M3, forged | 79NM | — | — | 40 | 14.4 |
| 72 | 110H6SV2MF, ESC | E3.29NK | 95 | — | — | — |
| | 160H5V7F5K5M3, forged | E3.29NK | — | — | 94 | 85 |

Note: Tests were performed in the temper rolling operations.

**Table 25.2.** Comparison of wear resistance of forged and cast-forged (in parentheses) ESR rolls of the 1680 rolling mill

| Roll material | Wear resistance of rolls (thousands of tons/mm of barrel wear) |
|---|---|
| 60H2SMF | 3.64 (7.2) |
| 9H2MF | 2.67 (2.7) |

**Fig. 25.2.** Electroslag castings for various components: roll with a diameter of 76 mm; forging die; hub and shaft of a marine diesel engine; coiler barrel sections; and compressor crankshaft section.

and thread cutting dies [5, 6]. Some of the parts produced by ESC and PMESC are shown in Fig. 25.2 and Fig. 25.3, respectively.

A comparison of wear resistance and the mechanical properties of the above parts produced by casting and forging is given in Tables 25.1 to 25.3. As is evident from Table 25.1, the wear resistance of forged rolls and of less highly alloyed ESCs is almost identical. Rolls for the multiple-roll U20 "Zundvig" mill were manufactured at the Starokramatorsk Manufacturing Plant, as well as at Swedish and German firms.

According to the data given in Table 25.2, the wear resistance of rolls made from forged castings with relatively little mechanical working is close to or even higher than that of forged rolls made of regular electric arc steel.

**Fig. 25.3.** Component preforms (tip, axle, support, casing, plate).

**Table 25.3.** Mechanical properties of components fabricated from electric arc steel forgings and from electroslag castings (ESC)

| Component type | Steel grade | $\sigma_t$ (MPa) | $\sigma_v$ (MPa) | $\delta$ (%) | $\psi$ (%) | KCV (Kj/m²) |
|---|---|---|---|---|---|---|
| ESC dies | 5HNM-Sh | 1,079 | 1,089 | 12.3 | — | — |
| Forgings satisfying SS 5950–78 | 5HNM | 750 | 750 | 10 | — | — |
| Marine diesel parts: | | | | | | |
| ESC shaft | 45-Sh steel | 280 | 570 | 28 | 43 | 620 |
| ESC hub | 45-Sh steel (45 steel)* | 330 (280) | 660 (540) | 28.5 (17) | 41 (38) | 400 (350) |
| ESC crankshaft throws of the MK-8 compressor | 34HNM-Sh (34HNM)** | 790 (500) | 930 (700) | 17.5 (14) | 54 (40) | 1,030 (500) |
| ESC segments of a rolling mill coiler barrel | 34HN3M-Sh (34HN3m)* | 760 (650) | 870 (800) | 18 (10) | 54 (30) | 1,180 (400) |
| ESC railroad car axles | 25 steel (25 steel)*** | 312 (220) | 477 (440) | 30 (20) | 58 (48) | 1,070 (500) |

Note: SS, State Standard; *forgings satisfying SS 8478–70; **requirements of the Gorki "Dvigatel revolyutsii" plant; ***forgings satisfying SS 8479–80.

The adoption of "NIIPTmash" developments in the field of electroslag technology (EST) at enterprises of the Ministry of Heavy Machinery (MHM) and other plants during the period of the XI five-year plan gave profits of about 1 million rubles.

The establishment of eight more ESC and PMESC shops at the MHM plants, with the help of the E.O. Paton Electric Welding Institute, and the intensive introduction of EST developments are being planned for the XII five-year period.

## References

1   I.K. Marchenko and E.I. Moshkevich, "Production of steel ingots by means of semicontinuous casting," Tehnika, Kiev, p.160 (1981)
2   I.K. Marchenko, "Advances in casting techniques for electroslag remelting and vacuum arc remelting of large electrodes," TsNIITmash *152*, pp. 34–36 (1980)
3   I.N. Ivanov and L.F. Vorobyova, "Economics of specialized metallurgy," Metallurgiya, Moscow, p.42 (1982)
4   V.I. Oleinichenko, G.A. Molodan, and G.D. Lyubchenko, "Improvements in techniques and equipment for increasing ESR steel quality," Application of New Materials and Alloys *12*, pp. 7–11 (1983)
5   G.A. Molodan, V.V. Bezhin, and A.V. Vishnevski, "Electroslag casting of intricate shapes," Electroslag Casting Technology *12*, pp. 3–4 (1983)
6   G.A. Molodan, V.K. Zabolotski, and L.N. Meshkova, "Contemporary production of multiple-roll mill rolls and future trends," Initial Material Production *3*, p. 34 (1984)

# 26 Production of Large Ingots by Proportional Electroslag Casting

Yu.V. Latash, A.E. Voronin, F.K. Biktagirov, R.G. Krutikov, V.B. Tynyankin, and Ya.M. Vasilyev

At present, electroslag technology (EST) is widely used in various branches of the metallurgical industry, such as in the production of large castings for the heavy machinery and power industry.

The technology of proportional electroslag casting (PESC) of large ingots originated at the E.O. Paton Electric Welding Institute. This is a method for increasing the size of ingots by sequential solidification of molten steel, which is poured in portions into the freezer mold. The surface of the molten steel is continuously heated by power supplied to consumable electrodes.

PESC is a relatively simple technique for increasing ingot size while maintaining steel quality. From analyses of the first PESC batches of various types of steel, it was determined that the transition zone between poured portions was not degraded. The first castings weighed between 5 and 15 tons and were produced from stainless steel, bearing, tool, roll, and constructional steels; this technique was later used to cast ingots[1] weighing 75 and 200 tons [1, 2].

With an appropriate choice of casting parameters, which depend on the weight of the ingot and the steel, it is possible to eliminate such defects as axial porosity, shrinkage pipes, and axial chemical inhomogeneity (V-segregation). Nonaxial inhomogeneity (L-segregation) is either absent or very low. These castings have a higher density and improved mechanical characteristics as compared to conventional steel castings. In addition, there is no deterioration of these characteristics in the axial zone of the ingot.

In PESC, the steel is intensely refined by the slag during pouring and during the long annealing period when other portions are being poured. The concentrations of sulfur and nonmetallic inclusions are reduced by a factor of 2 to 2.5. The oxide inclusion content (by volume) is about 0.004% to 0.006%; they are characterized by a uniform distribution throughout the ingot and are from 4 to 8 $\mu$m. The average content (by volume) of sulfides is 0.006% to 0.012%, and their concentration decreases radially from the surface and axially from the bottom to the top of the portion.

---

[1] Such 5-to 15-ton PESC ingots were produced at the Chelyabinsk Metallurgical Plant and the Electrostalsk Heavy Machinery Plant; 75- and 200-ton ingots were produced at the Kramatorsk "Energomashspetsstal" Plant.

---

Authors' affiliations: Yu.V. Latash, A.E. Voronin, F.K. Biktagirov, R.G. Krutikov, and V.B. Tynyankin, E.O. Paton Electric Welding Institute, Academy of Sciences, UkSSR; Ya.M. Vasilyev, MRA "TsNIITmash," USSR.

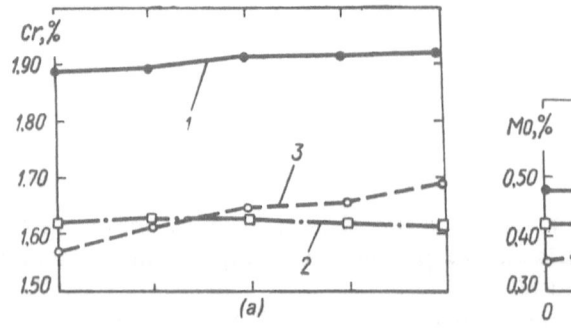
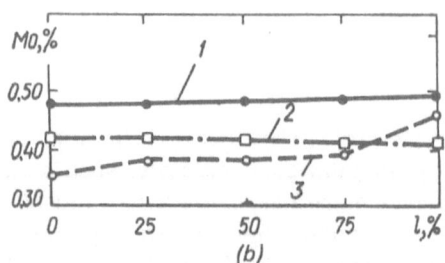

**Fig. 26.1.** Radial distribution of chromium (a) and molybdenum (b) in horizontal sections at a height of 60% of the ingot height ($l$, distance from ingot surface): (1) A 75-ton PESC ingot with a diameter of 1.8 m; (2) an 80-ton ESR ingot with a diameter of 2.3 m; (3) a 100-ton conventional ingot with a diameter of 2 m.

There is a small radial variation in distribution of elements with high segregation coefficients (sulfur, phosphorus, chromium, molybdenum, etc.), whereas conventional castings have significant axial and radial variations in the concentration of these elements. An example of the degree of macrosegregation in ingots obtained by various methods is given in Fig. 26.1, which shows the distribution of chromium and molybdenum in rotor steel ingots produced by conventional foundry methods, electroslag remelting (ESR) [3], and PESC. It is well known that the solidification of large volumes of molten steel (tens and hundreds of tons) is characterized by significant carbon segregation and the development of carbon-rich and carbon-poor zones in the bottom and the top of the casting, respectively. Carbon-rich zones do not occur in PESC castings except for a very low increase of carbon in radial sections of the last steel portions to solidify. For example, the maximum difference in carbon concentration in a 75-ton PESC rotor steel casting was 0.06%, whereas the carbon concentration in a similar conventional casting varied by more than 0.12%.

The development of segregation zones in steel castings depends to a large extent on processes that take place during solidification, to produce microsegregation or dendritic segregation. The dendritic inhomogeneity coefficient is an important characteristic of the casting. A comparison of chromium and molybdenum concentrations in the interdendritic regions of conventional, PESC, and ESR ingots is given in Fig. 26.2. It can be seen from the diagram that chromium and molybdenum concentrations in the central region of the conventional ingot increase radially; in PESC and ESR castings, the concentrations of these two elements are slightly increased in the direction of the advance of the solidification (front from the surface of the casting toward the center).

It should be mentioned that, even though the chromium concentration in the PESC casting is higher than in the conventional casting, the degree of segregation is, in fact, less because of differences in concentration of this element in the molten steel. This is supported by the results of calculations of the dendrite

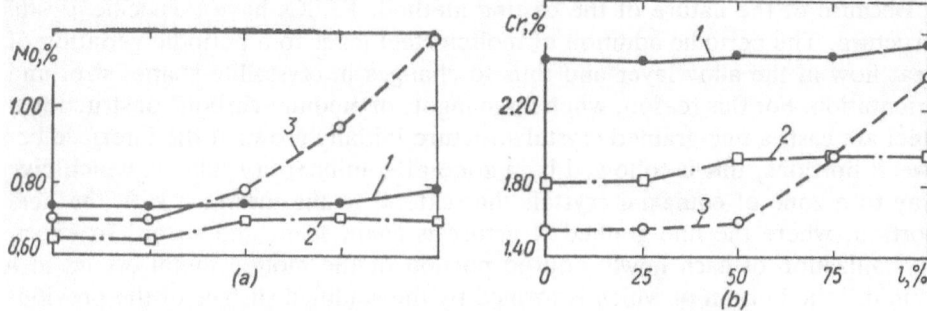

**Fig. 26.2.** Molybdenum (a) and chromium (b) concentration in the interdendritic zones of ingots produced by different methods (the curve number refers to that in Fig. 26.1).

segregation coefficient ($K_d$) given in Table 26.1. The coefficient $K_d$ for a given element is defined as the ratio of the concentration in the interdendritic zones to the initial concentration. The degree of segregation of PESC steel is close to that of ESR steel.

PESC steel has a relatively uniform distribution of alloying elements and impurities, a high density and homogeneity, a minimal degree of segregation, and a low level of shrinkage defects, all of which are due to favorable solidification conditions. One of the main advantages of this technique is that it is possible to control the electric power input to the slag and the feed rate of the molten alloy independently. By varying the weight of the alloy added, it is possible to maintain a relatively thin molten layer on top of the solidified steel; meanwhile, electroslag (ES) heating provides a high temperature gradient across this layer, which reduces the volume of the two-phase zone and the duration of its contact with the steel. For example, in casting a 75-ton PESC ingot of 30H steel with a diameter of 1.8 m, the thickness of the two-phase zone did not exceed 150 mm. For comparison, according to the results given in the literature [4], ESR of a 50-ton ingot with a diameter of 1.5 m produces coarse segregation defects, including noncenter line segregation, in those parts of the ingot where the distance between the liquidus and solidus isotherms exceeded 200 to 300 mm during remelting.

**Table 26.1.** Molybdenum and chromium dendrite segregation coefficients in ingots produced by different techniques

| Production technique | Segregation coefficients $(K_d^{Mo}/K_d^{Cr})$ | | |
|---|---|---|---|
| | Surface | One-half radius | Center |
| PESC | 1.5/1.15 | 1.65/1.25 | 1.85/1.30 |
| ESR | 1.4/1.10 | 1.58/1.20 | 1.80/1.27 |
| Conventional | 1.65/1.15 | 1.75/1.25 | 2.90/1.70 |

Because of the nature of the casting method, PESCs have a specific crystal structure. The periodic addition of molten steel leads to a periodic variation of heat flow in the alloy layer and thus to changes in crystallite shape, size, and orientation. For this reason, when large ingots of medium-carbon constructional steel are cast, a fine-grained crystal structure initially grows at the interface between portions; this is followed by a zone of columnar crystallites, which give way to a zone of equiaxial crystals that extend to the interface with the next portion, where the fine-grained structure is again seen, and so on. In effect, crystallization of each newly poured portion of the molten metal occurs in a "mold," the bottom of which is formed by the soldified surface of the previous portion.

Thus, in large PESCs, a fine-grained, columnar crystal structure develops not only in the vicinity of the surface, as in conventional castings, but also at interfaces between portions. However, in spite of the variation in crystal shape and size that takes place during solidification, the structure of PESCs is more homogeneous than that seen in conventionally produced castings, which have "frozen" crystals in the bottom and large coarse dendrites in the axial zone. As shown by the determination of dendrite size, the distance between dendrite axes along the cross section of PESCs is significantly smaller than in conventional castings. For example, the distance between the secondary axes of dendrites in a 75-ton PESC rotor steel casting increases from only 0.18 to 0.25 mm at the surface to 0.35 to 0.45 mm in the center, whereas in regular castings, the distance between secondary axes, especially in the center, is several times higher.

**Fig. 26.3.** Dendrite structure of the interface between portions in PESC steel.

The interface between portions exhibits a more dense dendrite sructure and a considerable reduction of the distances between the axes (Fig. 26.3). As a result, the steel in these regions in denser and has fewer nonmetallic inclusions and detrimental impurities and better mechanical characteristics.

As shown by operational tests, PESC provides favorable solidification conditions for molten steel and its refining by slag. The combination of modern melting and processing techniques that provide high-quality molten steel with PESC sharply reduces the incidence of segregation and shrinkage defects and is a promising method for producing of high-quality, large steel castings. Another advantage of PESC is that it can be used in relatively small capacity furnaces to produce castings weighing hundreds of tons. The choice of ES heating conditions of the casting head provides a final product yield of up to 90%.

## References

1   Yu.V. Latash, A.E. Voronin, and V.A. Nikolaev, "A new method for production of high quality large castings," Problems of Specialized Electrometallurgy 2, pp. 26–31 (1975)
2   Yu.V. Latash, A.E. Voronin, and V.A. Nikolaev, "Production of high quality large ingots by means of proportional electroslag casting," Steel 11, pp. 999–1002 (1975)
3   A. Choudhury, R. Jauch, and H. Lowenkamp, "Primarstuktur und Junenbeschaffenschmeltzverfahren hergesteller blocke miteinem Durchmesser von 2000 und 2300 mm," Stahl und Eisen 20, pp. 946–951 (1976)
4   T. Niimi, M. Miura, S. Matumoto, and A. Suzuki, "Analysis of large castings produced by means of ESR," Electroslag Remelting 6, pp. 308–322 (1975)

# 27 Large Hollow Castings of Quasi-Monolithic Reinforced Steel

B.E. Paton, B.I. Medovar, B.I. Shukstulski, V.Ya. Saenko,
A.D. Chepurnoi, V.V. Lapin, and V.V. Chernyh

To satisfy the current needs of industrial technology, it is often necessary to produce larger and larger steel products. Examples of large single components are the multiton rolls of the "5000" mill, turbine rotors used in nuclear power plants, and large engine shafts. And the forged shaft of a low-speed turbine rotor (1,530 rpm) used in a nuclear power plant with a capacity of 2 million KW is fabricated from a casting weighing between 400 and 500 tons. Thus, the problem of producing steel with a low content of nonmetallic inclusions and detrimental impurities that is more homogeneous is of great importance. As more new techniques of steel desulfurization, dephosphorization, reduction, and degassing have appeared, metallurgists now have the possibility of producing clean molten steel. The main problem is that forging large castings only partly eliminates casting defects. Also, larger forgings retain more casting defects, which are detrimental to the service characteristics of the product.

Electroslag remelting (ESR) was the only method used in the Soviet Union to produce high-quality castings weighing from 20 to 60 tons until 1980. The need for larger, high-quality castings (from 100 to 200 tons) made it necessary to develop new methods of processing molten and solidified steel.

The E.O. Paton Electric Welding Institute proposed a technique [1] for producing quasi-monolithic reinforced (QMR) steel with predetermined properties by using internal chills with a steel framework that is not melted during casting. The E.O. Paton Institute, the "Azovstal" Plant, and the MA "Zhdanovtyazhmash" jointly studied the use of internal chills that do not melt during casting, as well as internal chills that are fully melted in the casting process. The results of these studies were used in the design of production techniques for large, forged, hollow castings of QMR steel [2] (Fig. 27.1).

The quality of QMR steel, as determined by its mechanical characteristics, satisfies not only the requirements of State Standard (SS) 977-75 for castings, but also SS 8779-70 for forgings. This permits the direct use of hollow QMR castings for critical components without additional forging (Fig. 27.2).

We consider that the control of solidification of large quantities of molten alloy by the introduction of steel internal freezers into the mold, that is, the use of the QMR technique, offers great new possibilities for the production of heavy hollow castings and other large components for the needs of heavy industry.

Authors' affiliations: B.E. Paton, B.I. Medovar, B.I. Shukstulski, and V.Ya. Saenko, E.O. Paton Electric Welding Institute, Academy of Sciences, UkSSR; A.D. Chepurnoi and V.V. Lapin, MA "Zhdanovtyazhmash," USSR; V.V. Chernyh, Heavy Machinery Industry of the USSR, USSR.

**Fig. 27.1.**  A hollow, 120-ton, QMR steel ingot.

**Fig. 27.2.**  A high-pressure vessel casing fabricated from a hollow, 120-ton, QMR steel casting.

## References

1   B.E. Paton and B.I. Medovar, "On a new type of metallic material," report to the AS UkSSR, series A, No. 9, pp. 100–102 (1960)
2   L.B. Medovar, A.D. Chepurnoi, and B.I. Shukstulski, "The ingot," Znanie, Kiev, UkSSR, p. 48 (1986)

# 28 A New High-Strength Quasi-Monolothic Reinforced Steel for High-Tonnage BELAZ Dump Truck Beds Used in Mining

B.E. Paton, B.I. Medovar, V.Ya. Saenko, V.K. Postizhenko,
V.I. Us, L.B. Medovar, V.I. Moiseenko, A.D. Chepurnoi,
L.S. Shepotinnik, N.A. Gurov, G.Z. Gizatulin, and P.L. Mariev

The problem of increasing the service characteristics and reliability of a wide variety of equipment is closely connected with continuous improvements in steel plate quality. Until very recently, the only solution to this problem was in the use of electroslag remelted (ESR) low-alloy and alloy steel plate, which provides the highest strength available in combination with high toughness and ductility [6]. However, ESR steel is relatively expensive, which limits its application. Thus, research is being carried out in the Soviet Union and abroad to develop alternative methods that would be competitive with remelting technologies. All these technologies are based on inducing solidification of the molten steel.

Studies performed at the E.O. Paton Electric Welding Institute, in collaboration with many other institutes and plants, showed that it was possible to replace electroslag technology (EST), in some cases, with quasi-monolithic reinforced (QMR) technique casting in which internal freezer molds are used, for example, for the production of steel plate for critical welded products [2]. At present, QMR steel, which is much cheaper than ESR steel, is used in industry and in civil engineering [5]. Some of the largest consumers of QMR steel plate and sheet are motor vehicle manufacturers; QMR steel sheet is used for chassis parts of the "Zaporozhets," LuAZ, and other motor vehicles [1, 4], and QMR steel plate is used for fabricating beds for the BELAZ high tonnage dump truck [5].

The beds of large dump trucks used in mining operate under conditions of high impact and vibration loads, as well as abrasive wear. The worst service conditions are encountered in transporting highly abrasive ore with a hardness of 15 to 20 on the Protodyakonov scale and during ore loading at low ambient temperatures. The bed damage consists of dents, holes, and cracks caused by repeated impacts, and metal thinning, caused by abrasion. For example, after only 50 to 70 thousand km, the beds of 40-ton BELAZ-548A trucks were worn to 70% to 75% of their original thickness, and holes were worn through the bottom. The beds were fabricated of 22-mm 09G2S steel plate and carried ore with a hardness of 7 to 8 on the Protodyakonov scale. The situation is aggravated if the

Authors' affiliations: B.E. Paton, B.I. Medovar, V.Ya. Saenko, V.K. Postizhenko, and V.I. Us, E.O. Paton Electric Welding Institute, Academy of Sciences, UkSSR; L.B. Medovar, IPL, Academy of Sciences, UkSSR; V.I. Moiseenko, Academy of Sciences, BSSR; A.D. Chepurnoi, MA "Zhdanortyazhmash," USSR; L.S. Shepotinnik, The S. Ordzhonikidze "Azovstal" Metallurgical Plant, USSR; N.A. Gurov and G.E. Gizatulin, The Zhdanovsk Metallurgical Plant, USSR; P.L. Mariev, BELAZ, USSR.

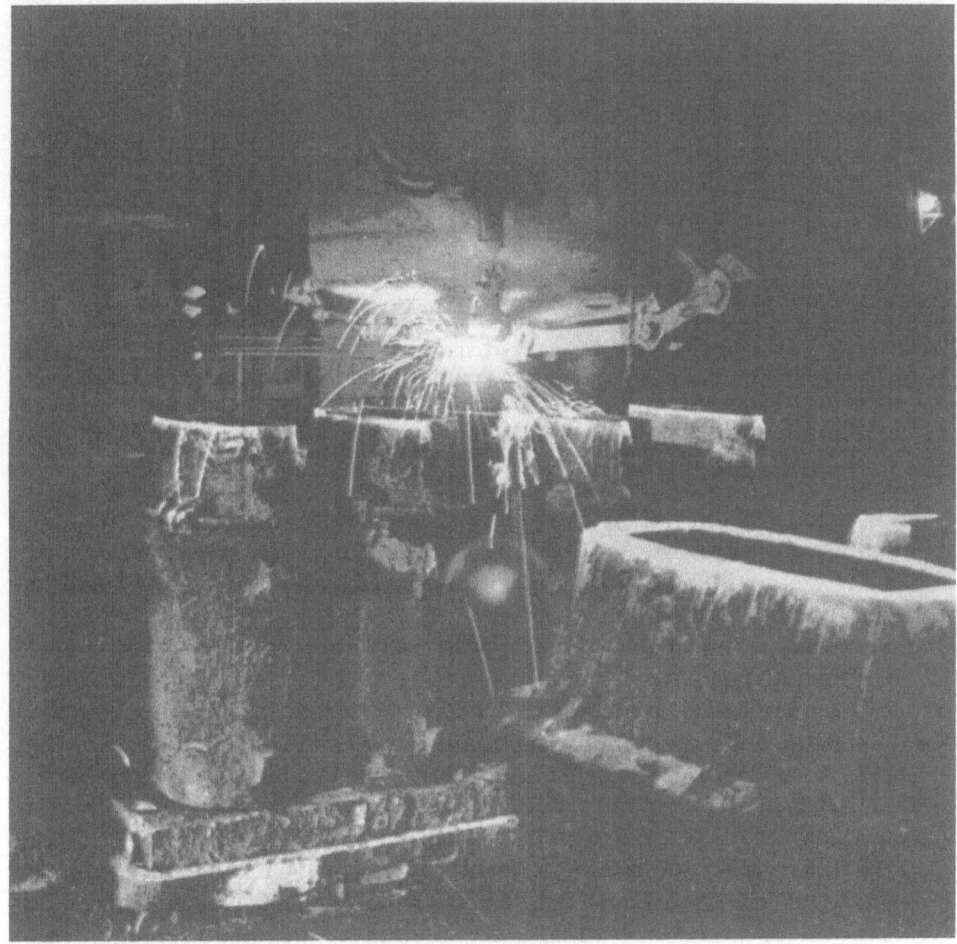

**Fig. 28.1.** A shop for producing quasi-monolithic reinforced steel.

vehicles are used at low temperatures, since even small dents may act as stress concentrations from which brittle cracks develop.

The bed wear resistance is not improved by low-alloy steels with better physical properties than 09G2S steel. Many practically new trucks that are used in mining are often out of service for bed repairs.

Analysis of specific physical and mechanical properties of QMR clad steel revealed that this new material has an increased resistance to wear under conditions prevalent in the mining industry.

In 1980 and 1981, steel 09G2SF-AKM, with improved properties, was tested as a replacement for 09G2S steel in the manufacture of beds for the 40-ton BELAZ truck.

For bed loading, a QMR steel that is usually used for gas pipes (TU14-1-3062-80) was chosen, as was a specialized reinforced, 09G2SF-QMR steel developed for use as bed bottoms. In order to increase the wear resistance of the bed,

**Table 28.1.** Properties of QMR steels used for manufacturing bed bottoms of BELAZ dump trucks

| Steel brand | $\sigma_t$ (MPa) | $\sigma_v$ (MPa) | $\delta_5$ (%) | $KCV_{60}$ (J/cm²) | $KCV_{-15}$ (J/cm²) |
|---|---|---|---|---|---|
| 09G2SF-AKM | $\dfrac{470–510}{490}$ | $\dfrac{570–600}{585}$ | $\dfrac{23–24}{28.5}$ | $\dfrac{10.7–18.8}{14.5}$ | $\dfrac{10.3–11.6}{13}$ |
| 09G2SF-AKM-U | $\dfrac{530–680}{(580)}$ | $\dfrac{620–710}{(665)}$ | $\dfrac{13–18.5}{(16)}$ | $\dfrac{4.1–11}{(75)}$ | $\dfrac{4.5–8.8}{(6.5)}$ |
| 22GSMTYu-AKM | $\dfrac{790–810}{(800)}$ | $\dfrac{900–950}{(925)}$ | $\dfrac{12.5–16.5}{(14.5)}$ | $\dfrac{4.1–7.4}{(5.5)}$ | — |

Note: Numbers above the line refer to the observed range of mechanical properties, below the line, to requirements for mechanical properties according to standard TU 14-1-3061-80. The numbers in parentheses are average values.

ingots were reinforced with inserts of type 70 high-alloy steel. Twenty-ton ingots of reinforced 09G2SF-AKM steel were subsequently rolled under controlled conditions into plates with an overall dimension of $17.5 \times 2{,}500 \times 10{,}500$ mm. The plate was cut into 5.2-m long sheets; their width corresponded to that of the BELAZ-548A truck bed.

The results of mechanical tests performed according to TU12-1-3061-80 showed that the high-strength reinforcing insert increased the wear resistance of 09G2SF-AKM-U steel, compared to 09G2SF-AKM pipe steel, without affecting the elongation and impact toughness coefficients (Table 28.1).

In 1981 about 200 BELAZ-540 dump trucks having beds of 09G2SF-AKM steel with a thickness of 17.5 mm were produced at the BELAZ. The beds of mass-produced BELAZ-548A trucks were made from 09G2S steel and had plate 22 mm thick. The performance of this trial batch of trucks demonstrated the advantages of the new bed material. The trucks have been in the field for several years in various mining operations and under different climatic conditions. The reliability of the truck beds is comparable to that of the thicker beds fabricated from 09G2S steel of mass-produced trucks. More specifically, the bed bottoms were in good condition, and no holes or breaks were observed. Thus, in trucks used for carrying coal and rock with a hardness of 4 to 5 on the Protodyakonov scale, the wear of the bed edge was no more than 1.5 mm after 100,000 km; in trucks carrying rock with a hardness of 12 to 14, the bed wear was no greater than 5 mm after 70,000 to 75,000 km, which is 20% to 30% lower than the amount of wear in standard, mass-produced trucks. The bed wear in vehicles carrying slate (hardness 7 to 8 points) was 1 mm after 80,000 to 100,000 km.

As a side benefit, the new bed material apparently increases the performance of tires on rough roads. It was observed that tire service life increased by 10% to 20% in trucks equipped with these new beds. This effect is not just due to a reduction in bed weight, it is also due to an increased resiliency and damping capacity of the bed, which reduce the impact loading of the wheels; as is well known, impact loads are two to three times higher than static loads, even at medium vehicle speeds.

The test results of dump trucks used in mining showed the effectiveness of

QMR steels for truck beds. The use of these steels reduces the weight of the bed, its initial cost as well as repair costs, and increases vehicle reliability. In addition, the low-alloy 09G2SF-AKM steel with a yield point of 470 to 510 MPa can be used for the production of beds of dump trucks with a load capacity of 40 tons.

For dump trucks with extremely high load capacities (up to 180 tons), which experience more intensive wear, steels with a higher strength and a yield point of 600 to 1,000 MPa are required. The wear resistance of conventional steels is increased primarily by increasing hardness, which is achieved by alloying with such expensive elements as molybdenum, niobium, and zirconium. Another important requirement on steels used in bed manufacturing is high weldability.

The highly uniform, plastic flow coefficient of QMR steel, which is achieved by metallurgical processing, allows for the production of steels with such characteristics as hardness, wear resistance, ductility, and fatigue and brittle fracture resistance. QMR steel is similar to ESR steel in that it is more homogeneous chemically and structurally, which increases the amount of plastic flow and, thus, the amount of energy necessary to fracture it. In other words, the use of QMR steels greatly improves such truck bed characteristics as toughness and fatigue limit, even if the properties of the QMR steel (tensile strength, yield point, elongation) are no different from that of the monolithic steels currently in use. In fact, QMR steels have better mechanical properties than do monolithic steels. For example, the cold brittle fracture limit of 09G2SF-AKM steel is the same as that of the more expensive and more highly alloyed 09G2FB monolithic steel. The endurance limit under axial cyclic loading with an asymmetry coefficient of 0.1 is 370 to 380 MPa, whereas the conventional 10HSND, which replaces the 09G2S steel, has an endurance limit of 250 MPa. This is very important because the amount of wear, according to fatigue wear theory, is inversely proportional to the product of the hardness and the plastic flow coefficient. Therefore, truck beds fabricated from QMR steel should exhibit lower wear and a higher fracture energy. In addition, the ability of QMR steel to prevent damage from spreading ought to decrease repair costs.

The increase of dump truck load capacity has sharply increased stresses on components, and therefore the beds of extremely high-capacity trucks are manufactured from a higher strength steel with a yield stress of over 600 MPa.

The bed bottoms of the mass-produced BELAZ-549 and BELAZ-75191 dump trucks with load capacities of 110 and 180 tons are fabricated from special heat treated, high-strength 14H2GMR steel. The severe service conditions of dump trucks used for carrying rock with a hardness of from 10 to 15 on the Protodyakonov scale intensify bed wear. For this reason, the beds of newly delivered trucks are additionally reinforced by the end users. Usually, sheets of St.3 steel with a total weight of 4 and 5 tons are welded on the bottom, front, and sides of the beds of BELAZ-549 and BELAZ-75191 trucks. This increases bed service life to 70,000 to 80,000 km.

A general factory overhaul consumes 7 to 8 tons of steel. The truck mileage before overhauls is 250,000 km, and 350,000 km at write-off. A total of four to five routine repairs and one major repair of the bed are required during the truck's service life. As a result of repair welding, the bed weight is increased, and the load capacity is decreased.

**Table 28.2.** Comparative characteristics of high-strength steels

| Steel type | Producing country | $\sigma_t$, MPa | $\sigma_v$, MPa | $\delta_5$, % | Impact toughness, (J/cm²) | $C_{EKV}$ | Concentration of alloying elements (%) in longitudinal samples | | | | | | | |
|---|---|---|---|---|---|---|---|---|---|---|---|---|---|---|
| | | | | | | | Mo | V | Cr | Ni | Cu | Nb | S | P |
| 14H2GMR | USSR | 600 | 700 | 14 | TU 14-1-2057-77 $KCU_{-40}=4$ | 0.985 | 0.5 | 0.03 | 1.7 | — | 0.3 | — | 0.005 | 0.016 |
| 14HGMDAFBRT | | 800 | 1000 | 12 | $KCU_{-40}=4$ $KCU_{-60}=3$ | 0.8735 | 0.3 | 0.2 | 1.3 | 1.3 | 0.9 | 0.006 | — | 0.0017 |
| 14H2GM | USSR | 600 | 700 | 16 | TU 14-1-2659-69 $KCU_{-40}=4$ | 0.873 | 0.5 | — | 1.7 | — | 0.3 | — | — | 0.016 |
| 14HGNMD | | 700 | 800–900 | 18 | $KCU_{-40}=4$ | 0.872 | 0.3 | 0.2 | 1.3 | 2.2 | 0.6 | 0.09 | — | 0.017 |
| WES135 steel | | | | | | | | | | | | | | |
| HW63 | Japan | 630 | 740–850 | — | $KCU_s=4.0$ | | | | | | | | | |
| HW70A* | | 730–770 | 810–830 | 20–23 | $KCU_{20}=15\div20$ | 0.73 | 0.5 | — | 1.05 | — | 0.3 | — | 0.003 | 0.007 |
| HW80 | | 800 | 880–1050 | — | $KCU_{-10}=2.8$ | | | | | | | | | |
| HW90 | | 900 | 970–1150 | — | $KCU_{-15}=2.8$ | | | | | | | | | |
| WEL-TEN100 | Japan | 900 | 970–1150 | 15 | $KCU_{-45}=15$ | 0.80 | 0.6 | 0.1 | 0.8 | 1.5 | 0.5 | — | — | 0.018 |
| N-A-KTRA100 | USA | 700 | 810–950 | 18 | $KCU_{-40}=2.6$ | 0.692 | 0.28 | — | 0.8 | — | — | — | 0.006 | |
| SS3100 | | 700 | 810–950 | 18 | $KCU_{-18}=6.9$ | 0.93 | 0.6 | — | 2.0 | — | 0.4 | — | 0.005 | |
| T-1 | | 700 | 810–950 | 18 | $KCU_{-46}=2.6$ | 0.74 | 0.65 | 0.08 | 0.65 | 1.0 | 0.5 | — | 0.006 | 0.02 |
| T-1 type A | | 700 | 810–950 | 18 | $KCU_{-40}=10.4$ | 0.63 | 0.25 | 0.08 | 0.65 | — | — | — | 0.006 | 0.021 |
| QT-35 | Britain | 570–590 | | 20 | $KCU_{40}=10.4$ | 0.7 | 0.5 | 0.12 | 1.0 | 1.2 | — | — | — | 0.085 |
| 22GSMT (AKM) | USSR | ≥800 | ≥900 | 14 | $KCU_{-60}\geqslant4$ | 0.68 | 0.25 | — | 0.3 | — | 0.3 | — | — | 0.001 |

A low-alloy, high-strength 22GSMTYu-AKM steel has been developed for the beds of high-tonnage dump trucks used in mining [7] (Table 28.1). This steel was produced for the first time in 1983 at the "Azovstal" Metallurgical Plant in a converter with a 350-ton capacity and was subsequently rolled into 18-mm plate on a 3600 rolling mill. The steel was temper hardened in a roller-quenching machine and was then used to manufacture beds of dump trucks with load capacities of 75, 110, and 180 tons. Commercial production techniques of this steel were developed at the Zhdanovski Metallurgical Plant in 1986. The steel is melted in a 160-ton oxygen converter and is bottom poured into type BS-24 molds, which are provided with reinforcing inserts. The QMR ingots are hot-rolled in the "1150" slabbing mill into slabs measuring $210 \times 1,400$ mm, which are finally rolled in the 3600 mill into plate with an overall dimension of $18 \times 2,300 \times 8,200$ mm.

When the 22GSMTYu-AKM steel is compared with other steels used for dump truck bed manufacturing that are produced in the Soviet Union and abroad, the superiority of the QMR steel (Table 28.2) is shown.

Service tests of BELAZ dump trucks with beds fabricated from the new steel, performed by the automobile industry jointly with the INDmash AS BSSR under various climatic conditions in the Soviet Union, gave satisfactory results: The bed bottom was in good condition, wear of the tailgate was low, no serious mechanical damage was observed, and no repair or bed reinforcement was necessary.

During these tests, the wear of beds fabricated from 22GSMTYu-AKM steel was 15% to 20% lower than that of mass-produced beds made of 14H2GMR steel. The use of 22GSMTYu-AKM steel prolongs the bed service life, reduce the costs of mining operations, labor, and bed repair, and increases the truck load capacity.

The use of high-strength 22GSMTYu-AKM steel for BELAZ-549V dump truck beds yields earnings of 13,000 to 14,000 rubles per unit.

# References

1  B.I. Medovar, V.Ya. Saenko, and V.I. Us "Quasi-monolithic reinforced steel and its use in automobile manufacturing," in "Problems of welding and specialized electrometallurgy," Abstracts of the All-Union Conference, Kiev, pp. 152–153 (1984)

2  B.E. Paton, B.I. Medovar, and V.Ya. Saenko, "Quasi-monolithic reinforced steel as an alternative to electroslag remelted steel," in "Electroslag technology," edited by B.E. Paton, Kiev, pp. 145–153 (1983)

3  B.I. Medovar, O.V. Berestenev, and M.S. Vysotski, "The use of the QMR steel for frames of nuclear power plants," Automobile Industry *10*, p. 17 (1985)

4  B.E. Paton, V.K. Postizhenko, and B.I. Medovar, "Use of QMR steel for manufacturing of automobile parts and assemblies," Problems of Specialized Electrometallurgy *2*, pp. 19–22 (1986)

5  V.Ya. Saenko, L.B. Medovar, and V.I. Us, "QMR steel as a new Constructional material," Znanie, Kiev, in the series "New Developments in Science, Technology, and Industry", No. 16 (1984)

6  B.E. Paton and B.I. Medovar, "Electroslag metal," Naukova dumka, Kiev, p. 680 (1981)

# 29 Quasi-Monolithic Reinforced Spring Steel

V.Ya. Saenko, V.I. Us, L.B. Medovar, M.S. Vysotski,
M.I. Gorbatsevich, V.I. Moiseenko, and V.I. Bondarkov

The service life of springs of vehicles used in mining and in the lumber industry, under severe conditions, is often lower than the life specified in their design. Since the weight of truck springs is 10% more than the chassis weight, an increase in spring reliability should not be obtained at the expense of increased weight.

Increasing the thickness of individual leaves and reducing the number of leaves reduce the total spring weight. However, the longer heating times that are necessary to process thicker leaves increases the decarburization of the steel, which sharply reduces the fatigue limit. The silicon spring steels that are widely used in automobile manufacture are susceptible to decarburization. Thus, thicker leaves are usually made from low-alloy or alloy steels, which increases the cost of the spring and makes modification of the heat-treating process necessary.

In view of these considerations, it was decided to study the service characteristics of springs made from quasi-monolithic reinforced (QMR) steels that were developed at the E.O. Paton Electric Welding Institute. These steels are produced by a process that is related to electroslag remelting (ESR), namely, freezing small volumes of alloy in an ingot mold or in a continuous-casting machine (CCM) freezer mold. For this purpose, macrofreezers, or reinforcements, fabricated from rolled stock, are introduced into the mold or the CCM prior to pouring in the molten steel. This reinforcement, which constitutes only 5% of the total ingot weight, does not mechanically reinforce the ingot as much as it allows rapid cooling and even solidification of the steel and eliminates all types of inhomogeneity. This steel, which is produced by a furnaceless technique, has fewer nonmetallic inclusions, and service characteristics close to those of ESR steel.

In the QMR technique, it is possible to control the degree of bonding between the reinforcement and the surrounding steel; this includes the melting of the insert if necessary. The relatively low weight of the reinforcement and the comparative simplicity of its use make it possible to lower the cost of QMR steels used in the manufacture of vehicle springs.

The new steels are characterized by more homogeneity and better resistance to nonuniform plastic flow prior to failure, which increases the strength of the decarburized layer.

---

Authors' affiliations: V.Ya. Saenko and V.I. Us, E.O. Paton Electric Welding Institute, Academy of Sciences, UkSSR; L.B. Medovar, IPL, Academy of Sciences, BSSR; M.S. Vysotski and M.I. Gorbatsevich, Minsk Automobile Plant (MAP), USSR; V.I. Moiseenko and V.I. Bondarkov, INDmash, Academy of Sciences, *Belorussian Soviet Socialist Republic* (BSSR).

**Table 29.1.** Chemical composition of QMR steel (%)

| Material | C | Mn | Si | S | P | Cr | Ni | Cu |
|---|---|---|---|---|---|---|---|---|
| Matrix | 0.60 | 0.75 | 1.64 | 0.019 | 0.010 | 0.12 | 0.06 | 0.08 |
| Reinforcement (framework) | 0.58 | 0.79 | 1.79 | 0.018 | 0.016 | 0.30 | 0.07 | 0.08 |

Extensive laboratory and commercial tests were performed at the INDmash AS BSSR and the MAP, as well as the E.O. Paton Institute and the Donetski Metallurgical Plant (DMP) to determine the properties of the new steel. A 5.5-ton, reinforced ingot of 60S2 steel was produced at the DMP. A welded reinforcing framework of 30-mm rolled stock of 60S2 steel was inserted into a bottom-poured mold. The chemical composition of the steel is shown in Table 29.1. The ingot produced by this method and a conventional witness-ingot poured from the same melting heat of alloy were subsequently rolled into a spring strip with a cross section of 90 × 12 mm.

A trial batch of rear springs #509-291211-B for the MAZ-509A truck and rear spring primary leaves for MAZ-type trucks was made from the QMR steel and from the monolithic steel of the witness-ingot at the Minsk Spring Plant. This batch of springs was tested according to the methodology of the Ministry of Automobile Industry of the USSR specified in "Spring Fatigue Limit Testing #37.001.013-73." Before the test, the spring load rate was measured; this is used to determine spring stiffness under static loading. The measured range of spring stiffness was 40 to 44 Pa for springs fabricated from QMR steel and 30 to 41 Pa for springs fabricated from conventionally produced steel. The hardness of leaf spring in both cases was higher than required by State Standard (SS) 9012-59 (363-444 HV). The average hardness of the QMR steel springs was 472 HV, and that of conventional springs, 497 HV. In order to analyze the quality of the material, the leaves that failed first during testing were selected for study. Deep etching of the surface did not reveal any defects. The thickness of the decarburized layer varied from 0.07 to over 0.14 mm.

In cyclic flexure tests, the number of cycles before failure occurred in the first leaf of QMR steel and of conventional steel was 155,000 to 180,000 and 60,000 to 145,000, respectively. Thus, it was determined that the service life of springs made of QMR 60S2 steel is 20% to 30% higher than that of springs made of monolithic steel. In addition, QMR springs were characterized by a lower scatter in their lifetime distribution. In November 1983, in parallel with the laboratory tests, field testing of leaves in the rear springs of MAZ-503A trucks was initiated at Minsk Automobile Plant #4. No spring failures occurred after 100,000 km. These results demonstrated the advantages of using QMR steel to produce steel used for automobile leaf springs, instead of the commonly used decarburization-susceptible steels.

# 30 Economic Aspects of the Use of Electroslag Castings To Produce Machine Parts at Kiev Factories

A.M. Brechak and S.V. Kryzhanovski

The main goals in the conservation of resources in manufacturing are to reduce the amount of scrap produced and to reuse existing scrap. Thus, a 50% reduction in the amount of scrap produced in machine manufacturing and metal working is equivalent to a 10% increase in steel production.

As is well known, steel is the major raw material used by the manufacturing industry. The development of many new technologies, in particular, electroslag casting (ESC), is aimed at reducing steel consumption. At present, ESC is widely used to produce machine components in manufacturing plants in the Kiev region. This is the result of a complex program, "Kievmetall." This program is based on the growth of advanced technologies, including electroslag crucible melting (ESCM), permanent mold electroslag casting (PMESC), and centrifugal electroslag casting (CESC).

In accordance with this program, electroslag (ES) techniques have been introduced in many Kiev manufacturing plants.

The ESCM technique was introduced with the goal of replacing conventional production techniques, particularly forging. The introduction of ESC has yielded earnings of 300 to 2,000 rubles/ton of castings. However, these values should not be regarded as an upper limit. Statistics from the Ministry of Light Industry (MLI) of the UkSSR show that earnings of 2,500 to 3,000 rubles/ton of castings can be achieved, depending on the efficiency of steel use. The use of ESC conserves steel because it produces castings with dimensions close to the final product, whereas forged components have high machining allowances (up to 500% of the final product weight). The steel consumption of ESC and forging production at the MA "Stroidormash" is shown in Table 30.1.

According to the data given in Table 30.1. the lowest steel consumption for forgings is 3.5, and the highest is 9.3; the corresponding values for ESC are 1.3 and 3.0, respectively, and the average steel consumption values for forgings and castings are 2 and 6, respectively. Thus, average steel consumption of ESC is one-third that of forging, which saves 2 to 6 tons of steel/ton of product. This consumption coefficient (see Table 30.2) accounts only for losses due to machining. The overall consumption coefficients, which take into account processing steps, are given in Table 30.2.

Thus, to produce 1 ton of components, of the type listed in Table 30.1, from forgings, 11.25 tons of steel are required, whereas ESC of these parts from rolled

---

Authors' affiliations: E.O. Paton Electric Welding Institute, Academy of Sciences, UkSSR.

**Table 30.1.** Steel consumption coefficients in the manufacture of machine parts from forgings and electroslag castings

| Component | Starting product | |
|---|---|---|
| | Forging | Electroslag casting |
| Epicycle gear | 3.5 | 1.3 |
| Lid | 6.2 | 2.4 |
| Half-sleeve | 6.0 | 2.0 |
| Bracket | 5.0 | 1.6 |
| Carrier | 9.3 | 3.0 |

stock requires 3.55 tons of steel and ESC using scrap requires 2.2 tons of new steel (3.2 times less). In addition, the utilization factor of metal (UFM) for forgings is only 0.09, which is 3.3 times lower than the industry average. The yearly saving of steel at the MA "Stroidormash" was 84.7 tons.

Tool carriages are currently produced by ESC at the Kiev Machine Tool Manufacturing Association (KMT MA). Formerly, this component, which weighs 170 kg, was fabricated from a forging weighing 380 kg, with a UFM of only 0.45. The ESC for this component weighs only 178.5 kg, which increases the UFM to 0.95 and yields a yearly saving of 1,235kg of steel/ton of finished product. This was made possible by selecting the best production technique and taking into account the special features of the new casting technology. The total saving at the KMT MA amounted to 25 tons of steel.

The introduction of ESC at the MA "Bolshevik" eliminated the need for forged welded products for some applications and reduced machining allowances by a factor of 10. In addition, this increased the UFM to 0.9 and saved 240 tons of steel in 1985.

ESC using consumable electrodes made from worn-out components and tools is regarded as the most effective conservation technique. For example, the PMESC technique is used at the MA "Stolovye Pribory" to produce rolling and stamping tools. Worn-out dies that are welded to make consumable electrodes

**Table 30.2.** Steel consumption coefficients for various processing methods [1, 2]

| Processing method | Starting product | | |
|---|---|---|---|
| | | Electroslag casting | |
| | Forging | Electrode from rolled stock | Electrode from scrap |
| Ingot production | 1.25 | 1.25 | — |
| Drop forging of ingots | 1.5 | — | — |
| Rolling | — | 1.29 | — |
| Electroslag casting | — | 1.1 | 1.1 |
| Machining | 6.0 | 2.0 | 2.0 |

are used as the starting material. This is a practically scrapless process in which starting material produced at the plant is used; it also decreases reliance on outside supplies of forgings and rolled stock of scarce steels (4H4M2VFS, H12S, etc.). According to plant records, the cost of a particular punching die produced by PMESC is 175.9 rubles, compared to 251.9 rubles for a conventionally produced die, and its service life is 1.2 times as long. The cost of L4117 coining dies produced by new versus conventional techniques is 89.9 and 137.7 rubles, respectively. The new technique increases the service life of this die 2.8 times. In addition to worn-out dies, steel scrap from other sources is remelted. The plant also manufactures tools for other factories.

The same strategy was adopted at the "Kievtraktorodetal" MA and other factories of the MLI UkSSR. The ESC shop of "Kievtraktorodetal" uses only tool steel scrap and worn-out dies for starting material. The ESR technique is particularly well suited to remelting high-speed steel from worn-out drills, milling cutters, etc.; it also increases the wear resistance of the tools produced. The adoption of this technique resulted in a savings of 600 tons of very scarce tool steel over a period of five years.

The PMESC method is used at plants of the MLI of the UkSSR to manufacture replacement parts that previously were imported. Worn-out machine parts and tools and rolled product scrap are used for the starting material. The quality of these ESCs is comparable to that of imported ones.

The production of castings with shapes close to the finished product significantly reduces the amount of machining necessary. As an example, at the MA "Strodormash," the necessary machining time was reduced by 1,492 man-hours, and by 4,498 man-hours at another factory, which freed up two mills costing 21,000 rubles. Furthermore, the machining operations that were needed did not require highly skilled personnel, such as lathe and milling-machine operators.

The reduced consumption of steel cuts down on the electric power used for casting, as well as that used in subsequent processing steps (forging, rolling, heat treating, etc.) and in machining.

In order to take full advantage of ESC for the manufacture of machine parts, casting should be used to replace production techniques that use starting materials inefficiently. The use of scrap steel for consumable electrodes substantially reduces the cost of the final product, which is determined mainly by the cost of the starting material.

In addition to the above factors, the reduction of equipment idle time also increases efficiency. At present, the average idle time of ESC installations is 50% to 60%, due to frequent changes in product type and also to low requirements for ESC parts at some plants. In order to utilize the equipment more fully, it should be matched to the product, and the possibility of supplying ESCs to other plants should be explored.

At present, there is a need for specialized regional centers to remelt scrap that is then used in the production of tools and machinery, both at the center and at participating factories.

One such center (belonging to the State Supply of the UkSSR) is now functioning in Kalinovka in the Kiev region. At this regional center, various types of components are produced from scrap by PMESC and CESC (shaping mill roll assemblies, forging dies, etc.). Future plans include using ESC in the production of large bearing races and recycling tool steels and guillotine shears and blades.

The experience of the plants of the Kiev region shows the effectiveness of ESC as a technique to conserve material and recycle scrap.

## References

1  K.M. Velikanov, "Calculation of the economic efficiency of new equipment," in "Mashino-stroenie," Leningrad, p. 430 (1975)
2  B.E. Paton and B.I. Medovar, "Electroslag metal," Naukova dumka, Kiev, p. 680 (1981)

144

**Photo 10.** An ingot produced by centrifugal electroslag casting.

PART 2

# Electroslag Technology Abroad

# 31 Modern Electroslag Remelting Technology and Its Future Development

V. Holzgruber

## Review

Over 30 years have passed since the first commercial electroslag remelting (ESR) installations were introduced at the Zaporozhie "Dneprospetsstal" Plant and, later, at Firth Sterling in McKeesport.[1] The Soviet installation was designed with a single stationary electrode; the American installation was the first to have replaceable electrodes.

Since that time, many more ESR installations have been put into operation and the production of ESR steel has grown to over 200,000 tons/year in Western countries (Fig. 31.1). At the beginning, the majority of the installations had the capability of remelting a single stationary electrode. At present, the majority of ESR steel is produced in installations having replaceable electrodes and a short movable freezer mold or a movable bottom plate (which is lowered as remelting proceeds).

Installations with a stationary freezer mold in which a single ingot is produced, using one consumable electrode, are well known. In such an arrangement, the ingot length is limited by the length of the freezer mold and the electrode. However, very long consumable electrodes are generally not economically feasible. This has led to the development of installations equipped with short movable freezer molds or with movable bottom plates,[2] with a capability for electrode replacement. These features make it possible to produce ESR ingots of a length that does not depend on the length of the freezer mold or the electrode because a single ingot can be obtained by the successive remelting of several electrodes. The first commercial installations of this type were put into operation at the Beler Company (Kapfenberg) and at the British Steel Corporation in Sheffield, in 1967. This technique is more advanced than the stationary mold process and is discussed in more detail later.

---

[1] The paper is printed in abridged form. For a more detailed discussion of ESR development, see "Electroslag Technology Abroad" by B.I. Medovar et al., Kiev, p. 320 (1982) [editor's note].

[2] This is based on the Soviet invention (patent #42969, "Method of Electroslag Remelting," Dec. 11, 1964), patented in Britain (patent #1103350, Nov. 16, 1965), Austria (patent #253711, Aug. 15, 1966), Italy (patent #653502, Dec. 1, 1966), and other countries [editor's note].

---

Author's affiliation: Inteco, Linz, Austria.

156                                                      V. Holzgruber

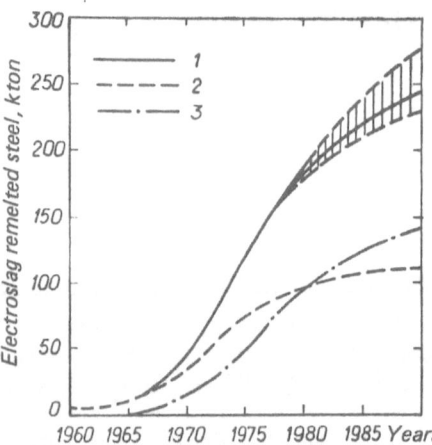

**Fig. 31.1.** Production of steel by electroslag remelting in the West, since 1966: (1) total; (2) with installations with movable freezer molds; and (3) with installations with interchangeable electrodes and a specially designed freezer mold.


The dependence of the slag consumption rate in ESR on the ingot length (for ingots with diameters from 400 to 1,000 mm) is given in Fig. 31.2. Since the slag consumption rate depends only on the ingot diameter, specific slag consumption is reduced for longer ingots. Slag consumption for ingots with a length from 1 to 1.5 m is 25 to 35 kg/ton, whereas consumption for ingots with an average length of 4 m is 10 kg/ton.

As shown in Fig. 31.3, the productivity of ESR installations depends on the ingot length. In order to maximize productivity, the length should be over 3 m. Furthermore, increases in the ingot diameter should be accompanied by corresponding increases in the length. This is because it takes more time to add flux, etc., into the mold when ingots with a larger diameter are produced. Interestingly, the length of the preparatory period (prior to remelting) does not depend on the ingot diameter.

The final product yield from ESR ingots depends on the ingot length because the lengths of head and tail croppings are fixed for a given diameter. In order to obtain a greater than 90% yield from ingots with diameters of 600 and 1,000 mm, the ingot length should be 2 m and over 4 m, respectively.

In addition to the above economic advantages, installations with interchangeable electrodes are not limited in the electrode: ingot diameter ratio. Very often, an electrode with a diameter larger than the freezer mold diameter is required to produce a single ingot of a given length when an electrode of a given length is used. In this case, numerous restrictions are placed on the process parameters (Fig. 31.4). Thus, it is not always possible to achieve such desired remelting conditions as solidification rate and depths of molten steel and slag layers. Usually, the ratio of electrode diameter to ingot diameter is between 0.4 and 0.5, and the upper limit is $0.6 \pm 0.1$; these values ensure minimum electric power consumption.

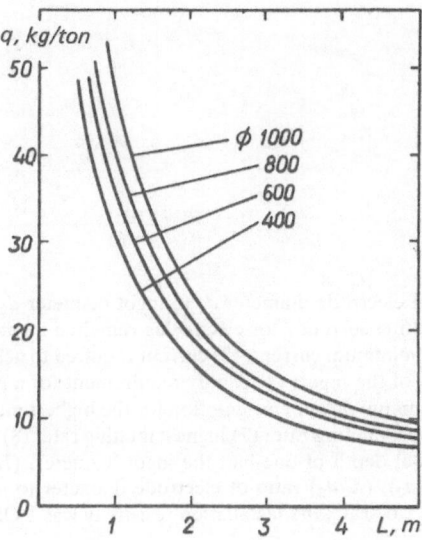

**Fig. 31.2.** Dependence of specific slag consumption $q$ on ingot length $L$ for ingots of various diameters.

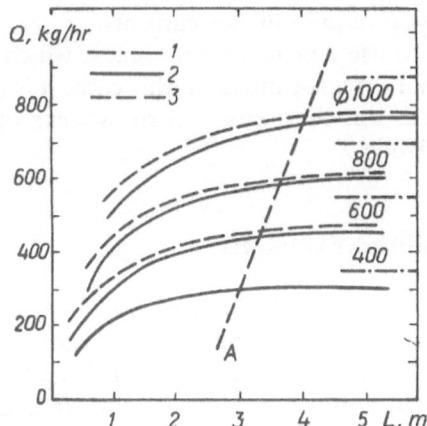

**Fig. 31.3.** Dependence of ESR installation productivity $Q$ on ingot length $L$ for ingots of various diameters: (1) melting rate; (2) ESR using a solid charge; and (3) ESR using a liquid charge. Curve for a melting rate that is 0.85 of the melting rate when a solid flux start is used (A).

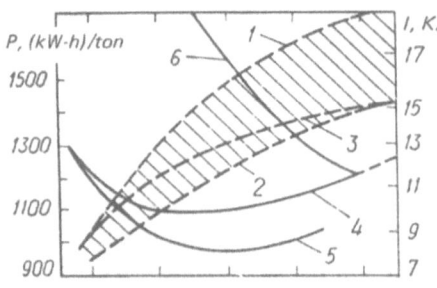 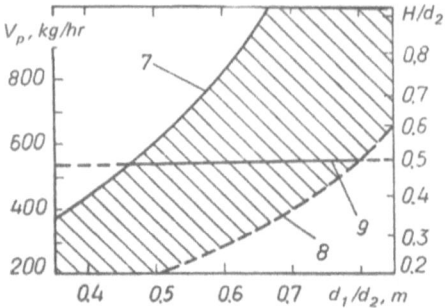

**Fig. 31.4.** Effect of the ratio of electrode diameter $d_1$ to ingot diameter $d_2$ on the melting rate $V_p$, power consumption $P$, and melting current $I$ for electroslag remelted ingots with a diameter of 600 mm: (1) maximum current; (2) minimum current; (3) current required to achieve a molten steel pool depth of one-half the diameter of the ingot; (4) energy requirement for a metal pool depth of one-half the ingot diameter; (5) minimum power consumption for the highest melting rate; (6) maximum power consumption for the lowest melting rate; (7) highest melting rate; (8) lowest melting rate; and (9) melting rate for a metal pool depth of one-half the ingot diameter. ($H/d_2$) ratio of metal pool depth ($H$) to ingot diameter ($d_2$); ($d_1/d_2$) ratio of electrode diameter to ingot diameter. The slag resistivity was 0.43 Ohm-cm at 1,600 °C; the installation resistivity was 1 Ohm-m.

ESR installations with replaceable electrodes are equipped with relatively long power cables that carry high currents, with a high inductive reactance. At commercial line frequencies, these installations function efficiently only up to currents of 20 to 25 kA. Therefore, low-frequency power sources were developed for installations that require higher currents.

Another solution is to use coaxial current leads, which are seen primarily in furnaces with a stationary freezer mold. In this case, it is possible to maintain a low furnace inductance, but there are still drawbacks of a stationary freezer mold and a single electrode.

## Modern ESR Installation Designs

In order to benefit from modern ESR technology, it was decided to develop a new installation that combines the advantages of both the above designs. The new installation (Fig. 31.5) consists of a short stationary freezer mold situated on the furnace platform and a movable bottom plate. In addition, the installation design permits replacement of the electrode to obtain the advantages that go with a capability of producing ingots of a length that does not depend on the freezer mold or the electrode length. In order to improve the performance of the secondary circuit of this furnace, a system of high-current leads was designed; it consists of water-cooled copper tubes connected by sliding contacts to high-current bars that run parallel to the electrode and the ingot. This arrangement was developed at the "Zhelezarna Ravne" steel mill (Yugoslavia) in an installation that produced ingots with a diameter of up to 1 m and a length of up to 6 m.

The installation, which was put into commercial production several years ago,

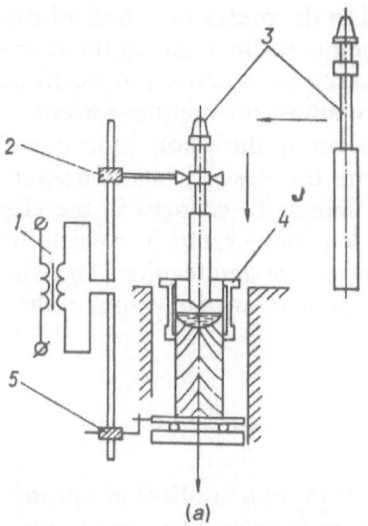

**Fig. 31.5.** (a) Diagram of the high-current circuit, including sliding contacts of an ESR installation equipped with an electrode interchange device and a movable bottom plate; (1) transformer; (2) electrode sliding contact; (3) consumable electrode; (4) stationary freezer mold; and (5) sliding contact for the movable bottom plate. (b) A photograph of the installation.

performed as expected, especially with respect to the inductive reactance, which was significantly reduced compared to that of existing installations with replaceable electrodes. The inductance of the new installation is only 40% the inductance of that of existing installations. As a result, installations of the new design can be used at commercial power line frequencies up to currents of 40 to 50 kA.

## Improved ESR Control Systems

In addition to an improved current supply system, the installation shown in Fig. 31.5 is also equipped with an improved control system to regulate electrode motion and to control the melting rate automatically.

The process control system consists of a computer that operates in real time, a weight sensor, and an electrode penetration regulator. The system regulates such ESR process parameters as the transformer voltage, melting current, and the depth of electrode penetration into the mold to correspond with the required melting rate for a given steel type, the cross section of the electrode and the freezer mold, and the composition of the slag. After initiation of the process, the actual melting rate is continuously compared to the preset rate, and adjustments are made to the slag layer voltage by a computer that controls the transformer voltage. In case of voltage fluctuations caused by variations of electrode depth, the electrode regulator can react rapidly to adjust the melting current.

This control system regulates the melting rate during the whole process and ensures that conditions are optimal for obtaining the desired ingot structure and surface finish. The system automatically responds to changes in the slag composition, the slag layer thickness, the electrode diameter, etc. The electrode weight sensor is used in automatic electrode replacement and feeding. The control system also provides a printout of the melting parameters and graphs of their variation.

## Directions for Future Development

Some trends in the further development of the ESR process are ESR in a protective atmosphere, ESR with a high gas pressure above the slag layer, and production of intricate shape castings by ESR.

### ESR in a Protective Atmosphere

ESR in a protective atmosphere with dry air introduced into the space between the electrode and the freezer mold wall has been used in many installations to reduce hydrogen incorporation and to prevent reduction of the slag's desulfurization ability. For remelting alloys that contain aluminum and titanium as reducing agents, an argon atmosphere is used.

Later, ESR in a protective atmosphere was used for low-alloy steels in order to reduce oxygen incorporation into the slag and, thus, to increase steel purity. Although ESR eliminates most nonmetallic inclusions, sometimes additional measures are required due to stricter steel quality requirements.

In order to meet very strict specifications, it is necessary to provide a protective atmosphere so that there will be no rejects because the steel is not sufficiently pure. Even the remelting of degassed electrodes is insufficient here because hydrogen enrichment of the steel in the bottom end of the ingot occurs, primarily because hydrogen is present in the slag. In contrast to ESR in air, ESR in a protective atmosphere completely eliminates hydrogen enrichment of the top end and the main part of the ingot. A protective atmosphere for the remelting electrodes of nondegassed steel also reduces the hydrogen concentration. This is evident from the data presented in Fig. 31.6, which shows the hydrogen concentration in a 760-mm diameter ingot 1 produced by remelting 30NSDV12 steel electrodes in air and in a protective atmosphere.

### ESR Under Increased Pressure

The solubility of a gas in molten steel depends on the pressure of the gas above the alloy surface. Gaseous nitrogen, which is soluble in iron and its alloys, is one

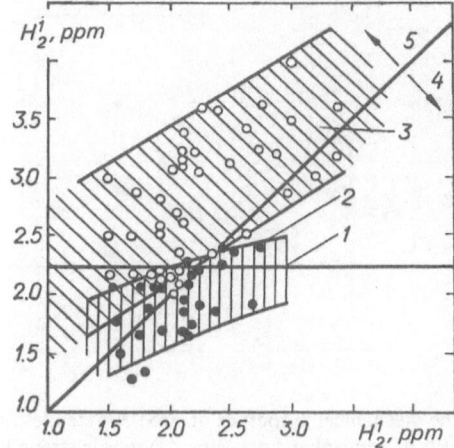

**Fig. 31.6.** Dependence of the hydrogen concentration in ingots remelted by ESR under various atmospheric conditions: (1) typical spread of values; (2) ESR in open air; (3) ESR in a protective atmosphere; (4) region of reduced hydrogen incorporation; and (5) region of hydrogen enrichment. ($H_2^i$) hydrogen concentration in the ingot, ppm; ($H_2^l$), hydrogen concentration in the electrode, ppm.

of the most active austenitizing elements. For this reason, melting and remelting processes under increased pressure are of great interest from the point of view of increasing nitrogen solubility in steel. Various methods have been proposed for obtaining alloys with a nitrogen concentration higher than the solubility limit at atmospheric pressure; these include induction melting, plasma-arc melting, and ESR under increased gas pressure.

Radiofrequency induction melting (RFIM) under increased pressure has been performed under laboratory conditions. In the USSR, plasma-arc melting under increased gas pressure has been put into commercial production, and the production of ingots weighing over 1 ton has been reported. Over the last several years, ESR under increased gas pressure has also been put into commercial production. Thus, an installation bult in West Germany in 1980 can produce ingots with a diameter up to 1 m and a weight up to 14 tons under a gas pressure of 4.2 MPa.

As mentioned, nitrogen is an austenitizing element. Dissolved nitrogen significantly increases the tensile strength of steel with an austenitic structure without appreciably affecting its ductility if nitride precipitation can be avoided. Earlier studies have shown that nitride precipitation in the majority of chromium–manganese austenitic steels can be prevented even at relatively high nitrogen concentrations.

Nitrogen is used as an alloying element to increase the yield point and tensile strength of austenitic stainless steels. The dependence of the mechanical properties—yield point and tensile strength (elongation)—of hardened AISI 304 stainless steel on nitrogen concentration is shown in Fig. 31.7. An increase in nitrogen concentration to 0.8%, which is possible only in ESR under increased

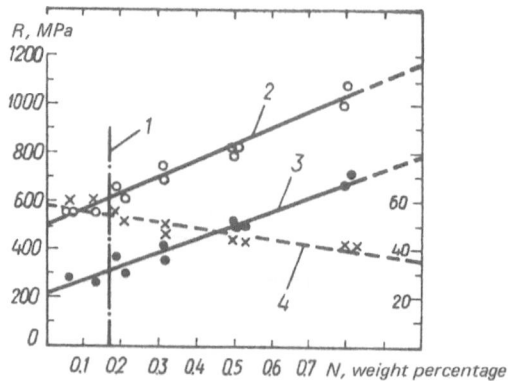

**Fig. 31.7.** Dependence of the mechanical properties of AISI 304 stainless steel on nitrogen concentration: (1) hydrogen solubility at atmospheric pressure; (2) tensile strength; (3) yield point; and (4) elongation.

gas pressure, significantly increases the yield strength from 200 to 600 MPa (200%). This method of increasing yield strength is of great interest, since the reduction in elongation is very small.

### Production of Intricate Shape Castings by ESR

The production of castings with a predetermined complex shape by ESR is growing in importance, and the production of cast rolls is of particular interest. The stages of roll production by ESR using an installation with a movable bottom plate and interchangeable electrodes are shown in Fig. 31.8. In casting this type of roll, slag volume should be increased at the transition from the roll neck to the thicker working body by the slow introduction of a fine-grained flux, but then, at the reverse transition at the other end of the roll, a portion of the slag must be removed. A special slag draw-off system has been developed for this purpose. The casting structure in the region of the transition from the lower roll neck to the body does not exhibit any defects, since the increased diameter and melting rate do not have any adverse effects on casting solidification or the shape of the molten metal pool. However, the melting rate must be controlled at the transition from the working body to the upper roll neck, so that no molten alloy will remain, in order to avoid secondary shrinkage defects.

About 30 roll castings with working body diameters of 630 and 740 mm and a roll neck diameter of 400 mm were obtained by the process briefly described in this chapter. Ultrasonic inspection of these rolls did not reveal any defects.

### Conclusions

This review of the development of ESR during the last 30 years has dwelt upon the reasons that led to the development of installations having the capability for

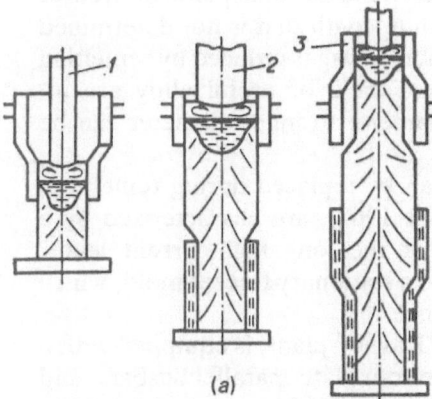

(a)

**Fig. 31.8.** (a) Diagram of the main step of roll production by ESR using an installation with replaceable electrodes and a movable bottom plate; (1) melting the lower roll neck (step 1); (2) melting the roll body (step 2); and (3) melting the upper roll neck (step 3). (b) Photograph of a portion of the roll casting.

electrode interchange and equipped with a movable bottom plate or a movable freezer mold. Electrode interchange and a movable bottom plate or freezer mold permit the production of ESR ingots with a length that is not determined by the electrode length. The specific consumption of slag is reduced by remelting long ingots, while installation productivity and yield of useful alloy are increased. In addition, the ratio of electrode diameter to ingot diameter can be varied widely.

ESR installations in which the electrodes can be replaced during remelting, and that have a short freezer mold (built since 1967), are characterized by a relatively high inductive reactance because of the long high-current leads, whereas installations with a single electrode and a stationary freezer mold, which have coaxial current leads, have a low inductance.

A new installation, built at the "Zhelezarna Ravne" plant, is equipped with a system for electrode interchange, a movable bottom plate, parallel busbars, and sliding high-current contact shoes in order to reduce the inductance of the furnace. In addition, the installation has a new control system that maintains a constant melting rate through a computer that regulates the transformer voltage and a special device that controls the depth of penetration of the electrode. This type of system can also be regarded as the basis for a completely automated ESR process for enclosed installations, as well as in furnaces with replaceable electrodes.

In addition, the chapter discusses some of the most important trends in ESR development such as ESR in a protective atmosphere, ESR under increased pressure, and the production of intricate shape castings by ESR. A reliable control system is a necessary precondition for realizing these processes. ESR in a protective atmosphere yields a low hydrogen concentration that satisfies the most severe requirements for steel purity; ESR under increased gas pressure permits the production of austenitic steels with a nitrogen concentration that far exceeds its solubility under atmospheric pressure. Since the presence of nitrogen in solid solution increases the tensile strength of austenitic steels, this process produces high-strength, non-magnetic alloys that are used for power-generating equipment. At present, ingots with a diameter of up to 1 m and a weight of up to 14 tons have been produced by ESR under increased gas pressure.

The production of rolls with a complex shape by ESR so that subsequent forging is not necessary, is very advantageous, economically. The chapter discusses equipment and techniques for roll production.

Finally, it should be mentioned that ESR was already highly developed some 10 years ago. In recent years, the main developments have been in improving existing installations and control systems to put into commercial production a number of specialized processes that, it is hoped, will be of great importance in the future.

# 32 Electroslag Remelting of High-Speed Steel in Hungary

## I. Sharvary

An electroslag remelting (ESR) laboratory [1–3] was established at the Budapest Institute of Ferrous Metallurgy (VASKUT) at the end of the 1960s. The equipment had the capability to produce ingots with a diameter of 115 mm and a weight of up to 40 kg.

The results, achieved with ESR under laboratory conditions, were comparable to published data. The properties of ESR RG high-speed steel, K1 and K13 tool steels, GOZ bearing steel, and steel for rolls were improved. The main drawbacks of the equipment were insufficient production capacity and small ingot size.

In order to increase the applicability of experiments and produce ingots of a size sufficient for tool manufacture, equipment with a pilot plant capacity was needed. Thus, a license for the USh-118 installation was purchased in the USSR, and the unit was put into operation in 1981.

## Tensile Strength Tests of High-Speed R9 Steel Produced by ESR Under Various Melting Conditions

High-speed steel belongs to the Lederburite group of steels. Eutectic carbides are present in the form of a network in conventionally produced steels, which decreases ductility in the as-cast state as well as in forgings.

ESR was chosen as the production method to improve steel characteristics because it is highly economical. An appropriate choice of remelting conditions permits the production of high-speed steel with the desired service characteristics.

There are numerous studies on the influence of remelting on the properties of steel, but very few published studies on interactions between basic remelting parameters and how they vary when different steels are remelted. A knowlege of these characteristics is very important for pratical applications.

The method of designed experiments was used for the study. Ten tons of R9 high-speed steel were produced by conventional melting (gas-fired furnace, permanent mold casting), and a portion was subsequently remelted using the parameters given in Table 32.1. The melting current, the voltage, the freezer mold filling factor, and the quantity of slag were varied in the course of ESR. In each case, a $CaF_2$-$CaO$-$Al_2O_3$ (60:20:20) slag, calcined at 800 °C was used. The

---

Author's affiliation: Budapest Institute of Ferrous Metallurgy, Budapest, Hungary.

**Table 32.1.** Parameters of electroslag remelting and different metallurgical technologies

| ESR process number | Remelting parameters | | | | | Metal working process | |
|---|---|---|---|---|---|---|---|
| | | | | Amount of slag | | Electroslag steel | |
| | I, KA | U, V | Mold filling factor | Percent ingot weight | Weight (kg) | Repeated forging (10 cycles) | Repeated forging (2.5 cycles) |
| 1 | 4.7 | 60 | 0.56 | 5 | 12.5 | B | A |
| 2 | 2.7 | 40 | 0.56 | 5 | 12.5 | D | C |
| 3 | 4.7 | 40 | 0.56 | 9 | 22.5 | G | F |
| 4 | 2.7 | 60 | 0.56 | 9 | 22.5 | K | H |
| 5 | 6.0 | 60 | 0.63 | 9 | 22.5 | M | L |
| 6 | 3.5 | 40 | 0.63 | 9 | 22.5 | V | V |
| 7 | 6.0 | 40 | 0.63 | 5 | 12.5 | T | S |
| 8 | 3.5 | 60 | 0.63 | 5 | 12.5 | P | N |

Note: The mold filling factor is defined as the ratio of the electrode diameter to the freezer mold diameter $f = D_e/D_f$. The letters $X$ and $Z$ denote conventionally produced steel (electric furnace melting, permanent mold casting).

average weight of the ingots was 250 kg; the average diameter was 222.5 mm. Both conventionally produced ingots and ESR ingots were then repeatedly forged (10 and 2.5 cycles); subsequently, all the forgings were simultaneously annealed (5 hours at 800 °C, and controlled cooling to 400 °C with a cooling rate of 25 °C/h). After complete cooling, the castings were cut into two pieces, one of which was used for tool manufacture and the other for mechanical test samples.

The standard characteristics that are taken as indicators of tool steel quality are chemical composition, hardness (after quenching or quenching and tempering), and carbide inclusion distribution and size. The determination of high-speed steel quality on the basis of carbide size is rather subjective, especially with respect to toughness.

Impact resistance (toughness) tests were performed on samples of hardened and tempered high-speed steel, but the results of such tests are not very reliable because of observed low values and a high scatter in the data. German scientists have developed a static bending test that is reliable for hardened and tempered steels, but it is not possible to estimate the effect of a particular metallurgical technique on high-speed steel quality from the test.

The goal of this investigation is to show the effect of the remelting technique on steel failure characteristics. In this respect, the nature of the heat treatment greatly affects eutectic carbide structure and steel toughness.

High-speed steel tools are conventionally used in the quenched and tempered state. Impact resistance tests, stationary bending resistance tests, and hardness tests are commonly performed on hardened and tempered alloy. In analyzing the relation of steel toughness to microstructure, one should take into account the fact that the microstructure in the quenched state is very brittle, and thus a fracture is characterized by a low crack propagation energy that is independent of propagation direction. Thus, neither eutectic carbide size nor distribution, nor metallurgical processing significantly affects the energy of fracture formation.

There are three stages of deformation fracture. In the course of static loading, a cavity initially forms in front of the crack or along the axis of the sample. Subsequent plastic deformation results in growth of the cavities and their coalescence with the crack tip. Cavity formation leads to the next two stages of deformation fracture. Thus, it is advantageous to provide conditions such that when cavities are formed in the eutectic carbides the surrounding material can absorb the resulting plastic deformation. This is where primary carbides play an important role.

It is obvious that testing tempered steel is the most appropriate way to study the problem. In order to ensure a reliable measurement of toughness, unnotched samples, identical to the conventional samples used for tensile tests, were studied.

The effect of variations in ESR parameters on fracture energy is shown in Table 32.2. The steel was repeatedly forged (10 cycles). The same type of tempering schedule was used for all the samples (5 hours at 820 °C followed by a slow cooling to 400 °C at a rate of 25 °C/h).

The letters denoting various metallurgical processes in Table 32.2 correspond to those explained in the note to Table 32.1. The samples marked X were fabricated from conventionally produced R9 high-speed steel, the standard for comparison. The other samples had the same chemical composition but were fabricated from steel obtained by ESR under different conditions.

The data given in Table 32.2 show the favorable effect of ESR; they also show that the value of $W_c$ calculated from tensile test results can be used to evaluate a given process or differences in structure.

**Table 32.2.** The effect of electroslag remelting on fracture energy ($W_c$) of high-speed steels

| Metallurgical process | $W_c$ (mm N)/mm³ calculated according to tensile strength tests | Relative value of $W_c$ (%) ($W_c X = 100\%$) |
| --- | --- | --- |
| X | 123.3 | 100.0 |
| B | 134.6 | 111.8 |
| D | 141.9 | 115.4 |
| G | 145.3 | 118.1 |
| K | 137.4 | 116.7 |
| M | 137.4 | 116.7 |
| P | 153.6 | 124.9 |
| T | 143.1 | 116.4 |
| V | 143.0 | 116.3 |

Note: $W_c$ is the energy of the plastic deformation fracture, calculated according to the following equation: $W_c = (R_{p0.2} + R_c) \ln d_o/d_c$, where $R_{p0.2}$ is 0.2 of the yield stress; $R_c$ is the ultimate tensile stress; $d_o$ is the original diameter; and $d_c$ is the diameter after fracture.

## Commercial Wear Tests of Tools Manufactured from High-Speed Steel Remelted on the USh-118 Installation

At present, wear resistance tests of tools fabricated from conventionally produced high-speed steel and ESR steel are still in the developmental stage. Thus, the results of two tests of tools fabricated of similar high-speed steels are given here. These tests were performed at the "RABA" plant (Dier). The wear resistance of milling cutters (74012E-GM) used for cutting internal teeth in BNC2 pinion gear blanks was determined (Table 32.3). All three types of tools manufactured from each of the steels listed in Table 32.3 were used on the same milling machine under the same working conditions.

It is evident from Table 32.4 that the chemical composition (which determines service properties) of R8 steel differs greatly from that of R9 steel (C and V) and of ASP30 steel (C, V, and Co).

It is clear from Table 32.3 that, in spite of the difference in chemical composition, the wear resistance of tools manufactured from R8 ESR steel is almost two times greater than that of tools fabricated from R9 steel and only 20% lower than that of tools manufactured from ASP30 steel. Thus, we can conclude that the wear resistance of high-speed ESR steel is at least two times greater than that of regular high-speed steel with the same chemical composition and is comparable to that of steel obtained by powder metallurgy, a technology that is currently regarded as an ideal production method.

The wear resistance of Klingelnberg-type gear cutting tools fabricated from different steels are compared in Table 32.5. The tools underwent the same type

**Table 32.3.** Wear resistance of gear milling cutters

| Steel type | Number of gears machined before regrinding is necessary | Total number of gears machined with one tool | Average wear of the cutter back surface (mm) |
|---|---|---|---|
| R9 | 20 | 454 | 1.1 |
| ASP30 | 23 | 1,150 | 0.5 |
| R8 | 23 | 958 | 0.6 |

Note: R9, conventionally produced high-speed steel from the Miskolc Metallurgical Plant (Hungary).

Editor's note: ASP30 is high-speed steel obtained by powder metallurgy; R8 is high-speed steel after ESR.

**Table 32.4.** Chemical composition of high-speed steels (%)

| Steel type | C | Cr | Mo | W | V | Co |
|---|---|---|---|---|---|---|
| R9 | 1.1 | 4.2 | 4.1 | 7.0 | 3.0 | 4.75 |
| ASP30 | 1.27 | 4.2 | 5.0 | 6.4 | 3.1 | 8.50 |
| R8 | 1.85 | 4.2 | 5.0 | 6.4 | 1.9 | 5.0 |

**Table 32.5.** Comparison of wear resistance of gear cutting tools fabricated from high-speed steel produced by different technologies

| Steel brand | Number of regrinding operations | Type of wear |
|---|---|---|
| ASP30 | 4 | Cold crumbling |
| R11 (MMP) regular | 5.8 | Wear + deformation of edge + crumbling |
| R11 (MMP) regular, after hardening in Germany | 4 | Crumbling |
| R11 (imported from Germany) | 11 | Wear + crumbling |
| R11 ESR (VASKUT) | 12.6 | Wear only |

Note: Chemical composition of R11 steel (%): C, 1.1; Cr, 4.2; Mo, 9.5; W, 1.5; V, 1.3; Co, 8.

of heat treatment and test procedure. Steel ASP30 shows the lowest wear resistance. The wear resistance of tools fabricated from ESR high-speed steel is at least 10% higher than that of tools of German high-speed steel and is two to three times greater than that of regular high-speed steel.

The results of these investigations show that ESR significantly improves the quality of high-speed steel, and especially Lederburite steel. As a result, it was decided to build two ESR installations to produce high-speed steel at the Miskolc Metallurgical Plant.

# References

1  E. Lendvai, "Electroslag remelting of steel," Kohasati Lopak *4* , pp. 160–163 (1965)
2  E. Lendvai, "Production of extra pure steels," Collection, VASKUT (1968)
3  E. Lendvai "Achievements in electroslag remelting in Hungary and possibilities of introduction into mass production," Kohasati Lapok *8* , pp. 349–356 (1972)

# 33 Electroslag Remelting of Specialized High-Alloy Steels

R. Schlatter and A. Bennani

The production of specialized steels by electroslag remelting (ESR) using a three-position installation made by the Birlek company was started at the Konie department of Deltasider in 1968. In 1983 this furnace was replaced by a larger universal ESR installation (Fig. 33.1) made by Leybold Heraeus, which can produce ingots with a diameter from 350 to 1,200 mm and a weight of up to 35 tons. The furnace has two external stations, for specialized freezer molds, and a main station, where electrodes are sequentially melted and the ingot is withdrawn from the freezer mold. The furnace is also equipped with two mechanically driven electrode holders, two transformers operating at line frequency, and an automatic control system. The main furnace characteristics are given below:

## Specifications of an ESR Furnace

| | |
|---|---|
| Furnace type | Three-position with two electrode holders |
| Freezer molds | Two stationary molds with a diameter of 850 mm and a 11-ton capacity; one mold with an ingot withdrawal system and a diameter of 1,200 mm and a 35-ton capacity (maximum calculated weight of the ingot, 44 tons) |
| Freezer mold length | Stationary, 3,200 mm; with ingot withdrawl, 1,100 mm |
| Ingot travel | Up to 5,100 mm |
| Electrode holders: | |
|   Load capacity | 15 tons |
|   Rotational drive | Hydraulic |
|   Linear | 3,900 mm |
|   Drive | Electromechanical, two motors |
|   Gear ratio | 3,000:1 |
|   Weighing system | Weight sensors, 0.03% precision |
|   Electrode positioning | Two-coordinate capability |
| Ingot | Shape: round, square, and rectangular in cross section |

---

Authors' affiliations: Deltasider, Italy.

**Fig. 33.1.** The Electroslag furnace at Konie.

| | |
|---|---|
| Dimensions | Diameter from 350 to 1,200 mm |
| Weight | From 1.2 to 35 tons |
| Length (with withdrawl) | Up to 5,000 mm |
| Electrode dimensions | Diameter, from 275 to 680 mm |
| Water consumption (recirculating water supply) | 220 m³/h (max.) |
| Emergency water supply | 35 m³/h |
| Power source | Two transformers with thyristor control |
| Circuit: | |
|   Primary | 5 kV, 2,400 kV-A |
|   Secondary | 80 V, 50 Hz |
| Maximum current | 25 kA, each transformer |
| | 36 kA, transformers in parallel |
| High-current circuit | Coaxial with three return current leads |

| Auxiliary equipment power supply | $3 \times 380$ V, 10 kV-A, 50 Hz |
|---|---|
| Control system voltage | 220 V, 50 Hz |
| Specific power consumption | 1,200 to 1,600 kW/ton |
| Computer control of melting | Control of start, melting rate, additional feeding, shape of solidification front, electrode replacement, electrode voltage, and feed rate |
| Number of melting programs in memory | 250 |
| Auxiliary equipment: Vibrating injector for alloying additives Air drying system | Capacity of $2 \times 15$ liters; consumption, 300–3,000 g/h; Consumption, 650 $m^3$/h under pressure of $5 \times 10^5$ Pa, dew point ($-18\,^{\circ}$C) |
| Electrode | By gas–oxygen burners |
| Flux melting furnace | Capacity of 300 liters (720 kg of slag), graphite electrodes, 600 kVA, 6 kA (max. current) |
| Exhaust ventilation | Flow, 2,000 $m^3$/h |
| Gas-scrubbing system | Additions of CaO, bag filters, dust concentration less than 50 mg/$m^3$ |

In this chapter, which is abridged from the original, ESR of Lederburite tool steels, ultra high-strength maraging steels, and heat- and corrosion-resistant alloys is described. It was found that ESR significantly improves the quality of tool steels, especially high-speed steels, because of the possibility of obtaining very large ingots with a low degree of mircro- and macrosegregation and a more uniform carbide distribution. The properties of maraging steels after ESR depend to a large extent on the chemical composition and the absence of nonmetallic inclusions. ESR of these steels is relatively complex and requires a variety of process operations, such as electrode preparation, provision of a controlled inert gas atmosphere, and reliable control of slag composition. Materials thus produced can be used in a number of demanding applications (except in the aerospace industry). However, further development of ESR is hampered due its complexity, compared to vacuum arc remelting (VAR).

ESR is widely used in the production of heat- and corrosion-resistant, precipitation-hardened alloys [1–4, 6]. In remelting larger ingots of these alloys, it is relatively easy to control segregation processes and the chemical composition, and ingot production is economical. As the concentration of the hardening elements titanium, aluminum, niobium increases, the size of the ingot fabricated from precipitation-hardening alloys has to be reduced because of reduced control over the chemical composition and segregation processes. In addition, production of these alloys is characterized by specific problems related to the necessity of strictly controlling active element concentration. When the ESR process is closely controlled, significant economic benefits can be had from the production of semifinished products from specialized iron- and nickel-based alloys.

## Integrated ESR Technology

The existing ESR technology is designed to produce round ingots with a diameter of 350 to 1,200 mm and a weight of up to 35 tons, and also ingots that are rectangular and square in cross section. Electrodes are obtained by up-hill casting into special molds that are fabricated from electric furnace steel subjected to AOD, VOD, or vacuum treatment in a ladle-furnace. To reduce shrinkage defects, hot tops are used. The electrodes are annealed before remelting, if necessary. Usually, the electrode surface is deburred; however, for some types of products the electrodes are ground after heat treatment to remove the casting skin or scale. The head and bottom ends of the ingot are cropped, if necessary.

The slag is usually prepared from standard raw materials. In some cases, a liquid charge is used, as are commercial premelted granulated slags. The furnace control system automatically performs start-up with a solid slag charge. For this purpose, exothermic mixtures that do not contaminate the alloy are used.

Usually, ESR is done at a constant melting rate. However, when large ingots are produced from alloys that are susceptible to segregation, the melting rate is continuously reduced to optimize solidification conditions along the ingot length. The computer control system allows for a flexible selection of melting conditions in order to provide a balance between the chemical and structural homogeneity of the ingot and the productivity of the process. Reducing agents are usually introduced in small portions during melting. Melting is terminated when a large part of the electrode is still unmelted. The reproducibility of termination and a high yield of useful alloy is achieved because of an accurate weight measurement system. After melting, the ingot is cooled for a period of time and then stripped either directly at the melting station or at a location outside the furnace. The ingot can be further cooled in the open air and then delivered to a heat-treating furnace.

## Electroslag Melting Technology

ESR can be regarded as a simultaneous refining and casting process, that allows for the control of alloy purity and crystalline structure. The main advantages of ESR of specialized steels are as follows: a high reproducible yield of alloy; high ingot homogeneity and high hot deformability; improved purity on both the micro- and macro-scales; and improved, uniform mechanical and processing properties.

The results of the application of ESR to three groups of materials are given below.

## Lederburite Tool Steels

Remelting of high-carbon, high-alloy steels is of particular interest because of
the improvement in mechanical properties, especially ductility and fatigue resistance, and a wider range of applications, due to the production of large ingots
with suppressed macro- and microsegregation and an improved carbide structure. Thus. ESR improves the quality of steels used for cold forging (AISI series
D), and high-speed steels (series M).

For Lederburite steels, the effect of the solidification rate on carbide structure, that is, the distribution and size of the carbide particles, depends to a large
extent on the ingot size. For a given ingot size, the optimum carbide distribution
in hot forged steel is achieved at a medium solidification rate. The size of the
carbide network cell at one-half the ingot radius decreases at an increasing solidification rate for small castings (300 mm diameter) and increases for larger
castings (over 500 mm diameter). For castings with a diameter of 400 mm and
over, the number of large carbides (over 20 $\mu$m) in the region between the center and one-half the radius is apparently slightly reduced or remains the same
with an increasing melting rate. However, the type and amount of hot working
will mainly determine the final carbide structure of the product (Fig. 33.2). This
is also true for ingots of cold forging tools steels of the D-2 and D-3 series with a
diameter of up to 500 mm. The thickness of the carbide segregation layers in M-2
steels increases from 40 $\mu$m, in ingots with a diameter of 300 mm to 75 $\mu$m in
ingots with a diameter of 500 mm; the average carbide size is 15 to 20 $\mu$m. The
practical melting rates used for these structure-sensitive Lederburite steels are
given in Fig. 33.3, along with the maximum ingot diameters permitted by the
process. When larger ingots of the same type of steel are melted, the melting
rate is evenly reduced by 20% to 30% up to termination. In stationary freezer
molds, the ingot length or weight is limited because of the need to maintain the
bottom end at a temperature that will prevent catastrophic cracking. In this respect, the remelting technique wherein the ingot is withdrawn through the bot-

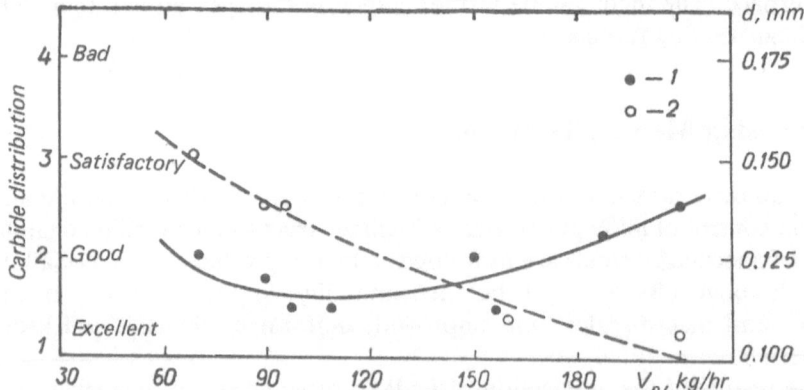

**Fig. 33.2.** The dependence of carbide distribution (1) and particle size ($d$) (2) on the solidification
rate ($V_p$) of ESR in high-speed steel ingots (ingot diameter, 300 mm).

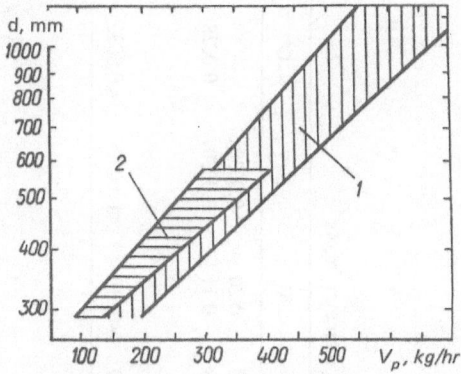

**Fig. 33.3.** Range of ingot solidification rates $(V_p)$ that yield ingots with an improved carbide structure from cold-forging Lederburite—series D steels (1), and high-speed series M steel (2). Ingot diameter $(d)$ in millimeters.

tom of the mold is more flexible. However, in all cases, heat-treatment of the ingot should be done carefully.

Successful ESR of large ingots of Lederburite tool and high-speed steels are related to improvements in the primary melting process and electrode casting techniques. When ternary slag compositions are selected, it should be kept in mind that the physical properties of the slag are more important in determining appropriate thermal and electrical conditions than are the slag's refining characteristics. However, a decrease in the concentration of nonmetallic inclusions is one of the main advantages of ESR steels, providing improved surface finish and increased fatigue resistance. These characteristics are the determining factors in many applications of tool and high-speed steels.

## Precipitation-Hardenign Steels

These steels belong to a group of ultra high-strength, high-ductility nickel steels. The characteristics of these steels are determined by the chemical composition and the concentrations of their metallic and nonmetallic inclusions. Thus, these steels are usually obtained from high-purity starting materials by double vacuum melting—radiofrequency induction melting (RFIM) followed by vacuum arc remelting (VAR)—or by VAR alone. Efforts to use ESR in this application have been unsuccessful because ESR does not limit the formation of nonmetallic inclusions and allow the necessary control over the amount of active elements (Ti, Al). The results of our research led to the manufacture of products from type 250 steel (1800 N/mm²) that are of an acceptable quality for "nonaerospace" applications.

An age-hardening steel containing 18% nickel was melted in an electric arc furnace, and subsequently processed by AOD. The charge materials were carefully selected. Cast electrodes thus obtained have a given chemical composition and a low concentration of detrimental impurities (Tables 33.1 and 33.2). Special attention was paid to the surface finish, the melting rate of the electrode, the slag composition, and control of the atmosphere.

Table 33.1. Chemical composition of age-hardening 250-type steel containing 18% nickel (1,800 N/mm$^2$) with 450-mm diameter ingots

| Sample | C | Si | Mn | P | S | Ni | Mo | Co | Ti | Al | N | O |
|---|---|---|---|---|---|---|---|---|---|---|---|---|
| Electrode | 0.010 | 0.07 | 0.04 | 0.004 | 0.003 | 18.75 | 5.00 | 8.60 | 0.60 | 0.20 | 0.0030 | — |
| ESR ingot (top) | 0.010 | 0.08 | 0.04 | 0.004 | 0.002 | 18.80 | 4.60 | 8.60 | 0.58 | 0.11 | 0.0030 | 0.0028 |
| *According to standard* | <0.012 | <0.10 | <0.10 | <0.006 | <0.004 | <18.5 | <4.90 | <8.50 | <0.50 | <0.10 | <0.0050 | <0.0030 |

Table 33.2. Mechanical properties of age-hardening 250-type steel containing 18% nickel (1,800 N/mm²)*

| Diameter of semifinished product (mm) | Sample orientation | $\sigma_{0.2}$ (N/mm²) | $\sigma_v$ (N/mm²) | $\delta$ (%) | $\psi$ (%) | Ku (J) | Fracture toughness, $K_{Ic}$ (N/mm$^{3/2}$) | Rockwell hardness, HRC |
|---|---|---|---|---|---|---|---|---|
| 105 | Longitudinal | 1,862 | 1,930 | 7.4 | 53 | 19 | 2,302 | 53 |
| 105 | Transverse | 1,860 | 1,930 | 6.2 | 40 | 14 | 2,220 | 53 |
| 57 | Longitudinal | 1,745 | 1,833 | 8.0 | 42 | 25 | 2,295 | 50 |
| | Minimum acceptable values | | | | | | | |
| | | 1,645 | 1,765 | 5 | 15 | 10 | 1,960 | — |

*Heat treatment: heated to 820 °C, air cooled, annealed at 480 °C for 3 hours, air cooled.

Since age-hardening steels are not resistant to heat, elevated temperatures can cause scale to form. This unstable oxide scale is soluble in the slag and interacts with active elements present in the slag. Thus, in order to remove the scale, cast electrodes should be ground or machined, and should also be protected from reoxidation in the course of remelting. It was determined that, for the best retention of titanium in the steel, protective paint coatings on the electrode surface are less effective than a low-oxygen, inert gas atmosphere above the slag layer. Here, argon is supplied by pipes welded to the electrode. Oxygen penetration into the upper part of the freezer mold is prevented by a close-fitting cap made from a fibrous refractory material. Titanium losses below 5% to 7% were achieved, dependancy primarily on two factors: regulation of the inert gas atmosphere and maintainance of proper slag conditions. Retention of active elements and alloy purity depend critically on the slag composition, purity, and remelting conditions. A ternary flux system can be used, consisting of 70% $CaF_2$, 15% CaO, and 15% $Al_2O_3$, with very low percentages of $SiO_2$, FeO, and C and with additions of up to 10% $TiO_2$. Too much $TiO_2$ increases the number of inclusions. In order to remove residual moisture, the flux was heated to 800 to 850 °C before use. The melting of 450-mm diameter ingots was initiated under a solid flux containing a small amount of exothermic powder. In the course of remelting, the molten slag was deoxidized by the continuous addition of granulated aluminum. For one melting run, titanium rods welded to the exterior of the electrode were used to deoxidize the slag.

The melting rate and melting conditions, particularly the depth of electrode penetration and the voltage, affect the composition, purity, and structure of the alloy and the ingot surface finish. Under the given conditions, the optimum melting rate was 250 to 270 kg/h. The chemical composition of the electrode and the ingot and the mechanical properties of the remelted alloy are given in Table 33.2. These data show that appropriate conditions and reliable process control give satisfactory results. On the whole, the properties of the ESR alloy are comparable to those of a VAR alloy; however, the electroslag alloy has a lower impact resistance and a lower isotropy.

## Heat- and Corrosion-Resistant Alloys

Heat-resistant iron-and nickel-based alloys can be divided into two groups, according to their metallurgical characteristics. Each group is characterized by specific requirements during ESR. For example, alloys with solid solutions that usually contain considerable amounts of volatile elements are easily remelted; age-hardening alloys, however, contain titanium and aluminum that oxidize readily and are lost due to interactions with the slag; this causes problems with alloy composition control.

The dependence of the diameter of the largest ingot that can be reliably obtained at present, by means of ESR, on the total amount of hardening elements is shown in Fig. 33.4.

In the case of simpler, precipitation-hardening alloys (AISI 330; Inconel 600™ and Incoloy 800™, (Alloys International Inco)), the main advantage of

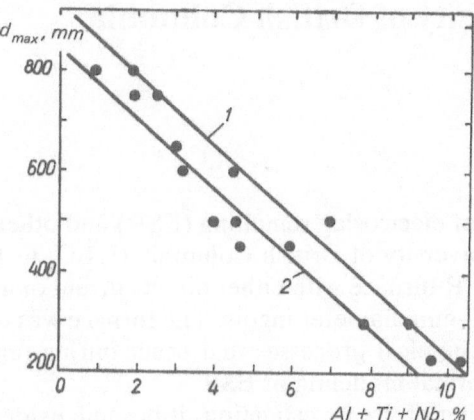

**Fig. 33.4.** Dependence of the maximum diameter ($d_{max}$) of ingots of heat-resistant alloys on the amounts of hardening elements. (1) ESR and (2) vacuum arc remelting.

ESR lies in its capability of producing large, high-quality ingots with high hot workability and increased yield. For example, Inconel 600 is remelted at an initial rate of about 700 kg/h; then the melting rate is reduced smoothly to 500 kg/h at the end of the process. In order to maintain constant amounts of titanium and aluminum, small amounts of $TiO_2$ are introduced into the slag as deoxidation with a NiMg/Al mixture continues.

Controlling the chemical composition is one of the main difficulties seen in ESR of age-hardening iron and nickel-based alloys that contain reactive elements and are microalloyed to improve their heat resistance [5]. Iron-based alloys pose the biggest problem. Many of the methods used for ESR of dispersion-hardening steels can be applied here. The behavior of active elements in the alloy is usually complicated and varies along the ingot length. For example, elements that are oxidized most during the starting period of the process are predominantly decreased at the ingot top-end. The titanium loss is low when the amount of the more reactive aluminum is relatively high, and there is no serious oxidation at the electrode surface that can disturb the oxide equilibrium in the slag. For an iron-based A-286 alloy with a high Ti:Al ratio (12. . . 16), titanium loss is 15% to 20%, even with careful slag preparation and proper remelting conditions, whereas for a Ni/Cr 80/20 alloy with a low Ti:Al ratio (1.5. . .1.8), titanium loss is less than 5%. Therefore, the ratio of titanium to aluminum in the alloy corresponds to the easily controlled $TiO_2$:$Al_2O_3$ ratio in the slag. However, one should take into account that a high concentration of $TiO_2$ reduces purity on a microscale. Reduced head croppings, a dense and smooth ingot surface, and improved hot forgeability of ESR alloys all contribute to the higher yield of semifinished products, compared to VAR alloys.

Thus, in order to obtain an ingot with a satisfactory structure, high-quality surface, and precisely controlled amounts of uniformly distributed elements, an accurate balance of numerous process variables should be maintained.

# 34 Research in the Field of Electroslag Remelting at the University of British Columbia

A. Mitchell

Research in the field of electroslag remelting (ESR) and other related processes was started at the University of British Columbia (UBC) in 1966 with the construction of a small ESR furnace with either direct current or a 60-Hz, AC power supply for melting 100-mm diameter ingots. The furnace was used to analyze the chemical and electrochemical processes that occur during remelting and to develop methods for thermal modeling of ESR.

At the very beginning of the investigation, it became evident that in order to obtain more information about the process it would be necessary to perform experiments using large commercial furnaces, which were not available at the university. Thus, a research program was initiated in which new methods would be developed using university equipment, and commercial testing of these methods would be done at outside plants when possible [1, 2].

A new, larger ESR furnace for producing 1-ton ingots with a diameter of 300 mm, using a 250 kVA, 60-Hz power supply, was built in 1976. This furnace can be used for electroslag welding (ESW) of large-diameter round stock—it has been used for welding steel stock with a thickness of 300 mm. This furnace can also be used for electroslag casting (ESC) of components weighing up to 500 kg.

## Analysis of ESR

Initial studies of electrochemical processes in ESR were aimed at describing the reactions taking place at the electrode end. It was determined that the Faraday reaction, $Fe(l) \rightleftarrows Fe^2 + 2e^-$, gives a good descriptive of the processes that occur at boundary surfaces in the course of ESR. We developed a mass transport model based on this concept that agrees with the observed variation in chemical composition even in large ESR ingots weighing up to 160 tons. The model also describes the formation of oxide inclusions and the deoxidizing processes that occur during ESR.

The thermal behavior of ESR is described using a two-dimensional model of the steady-state process [3, 4]. The model can be used to predict the structure of complex alloys, such as tool steels and superalloys. At present, the model is used in ESR control systems in Canada, the United States, France, England, Brazil, Japan, and Australia.

---

Author's affiliation: University of British Columbia, Vancouver, British Columbia, Canada.

**Fig. 34.1.** A casting of the shut-off valvè body.

**Fig. 34.2.** An aluminum freezer mold for electroslag casting of drive gear blanks.

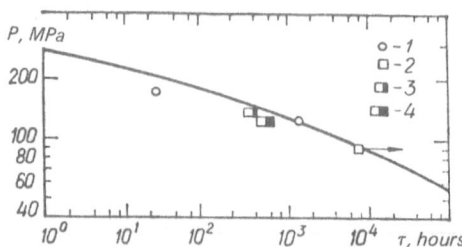

**Fig. 34.3.** Test results for the valve body casting (CF8M stainless steel, test temperature 649 °C): (1) axial stress; (2) transverse stress; (3) ingot stress prior to heat treating; (4) ingot stress after tempering; and ($\tau$) time to failure.

**Fig. 34.4.** The installation for the electroslag additional feeding of ingots

## Studies of the ESC Process

The research program on ESC of complex shaped components was started in 1976. For this purpose, three types of products were selected: a high pressure (1,360 atm.) stainless steel shut-off valve body, a ball mill drive gear of SAE 4340 steel, and an inlet pipe of A516 steel. For reasons of economy, we developed a casting technique using aluminum freezer molds (Figs. 34.1 and 34.2). Analysis of the mechanical properties was performed in cooperation with commercial plants and included both research work and conventional commercial nondestructive tests of these components [5, 6]. As a result, several commercial ESC plants were built in North America. One of the commercial test results is shown in Fig. 34.3.

## ESW and Additional Feeding

The installation at UBC was also used for developing heat transfer models describing ESW of large cross-section, semifinished products and electroslag (ES) additional feeding of large ingots. These models were used for building commercial plants (Fig. 34.4). Recently, the world's largest installation for ES additional feeding of ingots based on these models was introduced into industrial production in the United States.

The studies of ESR at UBC are directed toward improving production techniques for high-quality carbon and alloy steels. All the problems of current commercial production were taken into account. We expect that research in this field will continue, since industrial needs for high-quality steel are increasing.

# 35 A 200-Ton Electroslag Furnace in the People's Republic of China

D. Dzue, L. Haihon, S. Shuie, D. Dzyanhe, and L. Chuanlin

The rapid development of power engineering has increased the need for high-capacity and high-quality generators and turbines. After investigating modern methods of producing large ingots for forging, we have chosen the electroslag remelting (ESR) technique, since it yields high-quality ingots.

In their beginning of the 1960s, a three-phase furnace was built at the Shanghai Heavy Machinery Plant. Its design was based on a small electric arc furnace. The first ESR ingot, weighing 100 tons, was produced in 1965. During the period from 1965 to 1972, 48 ingots weighing from 100 to 500 tons were produced at this installation.

Since there was a need in industry for ingots weighing over 200 tons, large furnaces have been studied since 1973, and in 1979, a 200-ton electroslag furnace was built at the Shanghai Heavy Machinery Plant (Fig. 35.1). The first ingot, weighing 89 tons, was produced in April 1980. This chapter describes technical characteristics of the furnace and presents analyses of the ingot quality, in particular, 24 ingots weighing over 100 tons and 2 ingots weighing 205 tons. Most of the ingots were forged, some of which were analyzed for quality.

As an example, we discuss the quality of an ingot weighing 127 tons (Fig. 35.2). The surface of the ingot is sufficiently smooth so that no conditioning before forging was required. The chemical composition of the ingot at the locations shown in Fig. 35.3 is given in Table 35.1. The electrode composition is given in Table 35.2. The distribution of some elements along the ingot cross section is shown in Table 35.3. These data show that the ingot has a homogeneous structure and does not contain the macrosegregation that is typical of conventional steel production technologies.

Sulfur prints of the transverse and longitudinal templates were made. Macrosegregation of sulfur was not observed. The oxide and sulfide concentrations were very low (Table 35.4), and large inclusions were not observed. The purity of the steel shows that ESR is sufficient to limit the presence of nonmetallic inclusions.

Mechanical test samples were cut from forgings after heat treatment (Table 35.5 and 35.6). These data show that ESR can produce ingots with excellent

---

This paper was presented at the VIII International Conference on Vacuum Metallurgy, Linz, Austria, September 30 to October 4, 1985.
Authors' affiliations: D.Dzue, L. Haihon, and S. Shuie, Peking University of Ferrous Metallurgy, Peking, People's Republic of China; D. Dzyanhe and L. Chuanlin, the Shanghai Heavy Machinery Plant, Shanghai, People's Republic of China.

Fig. 35.1. Overall view of the 200-ton electroslag furnace.

mechanical properties, especially ductility. Ultrasonic inspection of parts after preliminary working did not reveal and defects over 2 mm in diameter.

The design of the 200-ton ESR furnace takes into account the experience gained in the use of the 100-ton furnace and numerous experimental results. Three single-phase transformers on a three-phase line are used to provide a balanced load, and the supply voltage can be changed under load. Six electrodes

Fig. 35.2 An 127-ton electroslag ingot.

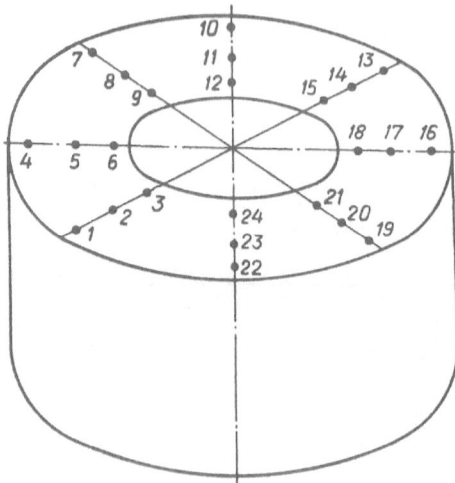

**Fig. 35.3.** Diagram of sample selection across ingot sections. Sample selection points, 1 through 24.

with a diameter of 500 mm are fixed in three electrode holders and are simultaneously remelted in a water-cooled, copper freezer mold. The design permits electrode replacement in the course of melting. The use of six electrodes provides uniform heat distribution throughout the molten metal pool. The electrode holder arms can be extended to position the electrode in the center of the freezer mold. This position is used in the production of smaller ingots using a single-phase circuit. The furnace has bifilar current leads, and its power factor is between 0.87 and 0.96. For melting and refining flux, a flux melting furnace was constructed. The use of a liquid flux charge and equipment for air drying prevents hydrogen enrichment of the steel during remelting. The freezer mold diameter is 2,800 mm. Withdrawal of the ingot from the mold as remelting proceeds makes possible the production of ingots weighing up to 240 tons. The furnace is equipped with a gas scrubbing system that provides a 99% purification of the exhaust. The furnace is also supplied with a backup water-cooling system that prevents process interruptions.

Up to the fall of 1985, 29 large ingots were produced. Analysis of the ingots confirmed the high quality of the steel.

The 200-ton electroslag furnace was developed jointly by the Peking University of Ferrous Metallurgy and the Shanghai Heavy Machinery Factory, and installed in Shanghai. All the 29 ingots produced in this furnace had high purity, improved mechanical properties, and a homogeneous chemical structure. Commercial use showed that the furnace has a satisfactory productivity and that ESR will play an important role in the production of large forgings.

**Table 35.1.** Chemical composition of the ingot (%)

| Sample selection point | C | Mn | Si | S | P | Cr | Ni | Cu | Mo | Nb | Co |
|---|---|---|---|---|---|---|---|---|---|---|---|
| Standard Specification | 0.17–0.23 | 1.20–1.60 | 0.15–0.30 | 0.015 | 0.012 | 0.25 | 0.60–0.90 | 0.05 | 0.45–0.65 | 0.02–0.06 | 0.02 |
| 1 | 0.23 | 1.30 | 0.24 | 0.003 | 0.008 | 0.03 | 0.93 | 0.028 | 0.56 | 0.050 | 0.006 |
| 2 | 0.20 | 1.46 | 0.23 | 0.003 | 0.009 | 0.03 | 0.89 | 0.029 | 0.57 | 0.049 | 0.006 |
| 3 | 0.19 | 1.48 | 0.29 | 0.003 | 0.009 | 0.03 | 0.88 | 0.028 | 0.57 | 0.047 | 0.006 |
| 4 | 0.20 | 1.47 | 0.29 | 0.003 | 0.007 | 0.03 | 0.87 | 0.029 | 0.56 | 0.048 | 0.006 |
| 5 | 0.19 | 1.48 | 0.30 | 0.003 | 0.009 | 0.03 | 0.88 | 0.024 | 0.55 | 0.047 | 0.006 |
| 6 | 0.20 | 1.50 | 0.30 | 0.003 | 0.013 | 0.03 | 0.88 | 0.024 | 0.56 | 0.046 | 0.006 |
| 7 | 0.19 | 1.50 | 0.30 | 0.003 | 0.009 | 0.03 | 0.87 | 0.026 | 0.56 | 0.046 | 0.006 |
| 8 | 0.20 | 1.49 | 0.30 | 0.003 | 0.008 | 0.03 | 0.86 | 0.024 | 0.54 | 0.049 | 0.006 |
| 9 | 0.20 | 1.50 | 0.31 | 0.003 | 0.013 | 0.03 | 0.88 | 0.030 | 0.58 | 0.047 | 0.006 |
| 10 | 0.20 | 1.52 | 0.27 | 0.003 | 0.012 | 0.03 | 0.88 | 0.030 | 0.56 | 0.048 | 0.006 |
| 11 | 0.22 | 1.50 | 0.25 | 0.003 | 0.011 | 0.03 | 0.89 | 0.032 | 0.56 | 0.047 | 0.006 |
| 12 | 0.20 | 1.50 | 0.23 | 0.003 | 0.011 | 0.03 | 0.89 | 0.030 | 0.56 | 0.050 | 0.006 |
| 13 | 0.22 | 1.48 | 0.29 | 0.003 | 0.012 | 0.03 | 0.87 | 0.031 | 0.57 | 0.050 | 0.006 |
| 14 | 0.21 | 1.48 | 0.30 | 0.003 | 0.009 | 0.03 | 0.86 | 0.030 | 0.56 | 0.046 | 0.006 |
| 15 | 0.20 | 1.48 | 0.31 | 0.003 | 0.006 | 0.03 | 0.86 | 0.030 | 0.57 | 0.048 | 0.006 |
| 16 | 0.21 | 1.48 | 0.29 | 0.003 | 0.009 | 0.03 | 0.87 | 0.030 | 0.57 | 0.046 | 0.006 |
| 17 | 0.22 | 1.50 | 0.29 | 0.003 | 0.008 | 0.03 | 0.88 | 0.030 | 0.58 | 0.047 | 0.006 |
| 18 | 0.19 | 1.46 | 0.28 | 0.003 | 0.010 | 0.03 | 0.85 | 0.030 | 0.55 | 0.049 | 0.006 |
| 19 | 0.22 | 1.52 | 0.24 | 0.003 | 0.012 | 0.03 | 0.88 | 0.030 | 0.59 | 0.048 | 0.006 |
| 20 | 0.22 | 1.48 | 0.24 | 0.003 | 0.013 | 0.03 | 0.87 | 0.031 | 0.55 | 0.049 | 0.006 |
| 21 | 0.20 | 1.45 | 0.26 | 0.003 | 0.011 | 0.03 | 0.91 | 0.032 | 0.55 | 0.048 | 0.006 |
| 22 | 0.18 | 1.50 | 0.30 | 0.003 | 0.011 | 0.03 | 0.91 | 0.031 | 0.55 | 0.049 | 0.006 |
| 23 | 0.22 | 1.50 | 0.25 | 0.003 | 0.010 | 0.03 | 0.89 | 0.031 | 0.56 | 0.046 | 0.006 |
| 24 | 0.23 | 1.50 | 0.21 | 0.003 | 0.010 | 0.03 | 0.91 | 0.032 | 0.56 | 0.046 | 0.006 |

**Table 35.2.** Chemical composition of the electrodes (%)

| Melting run number | C | Mn | Si | S | P | Cr | Ni | Cu | Mo | Nb | Co |
|---|---|---|---|---|---|---|---|---|---|---|---|
| 81A191 | 0.21 | 1.47 | 0.22 | 0.014 | 0.010 | 0.11 | 0.92 | 0.05 | 0.49 | 0.030 | 0.003 |
| 81A193 | 0.22 | 1.47 | 0.21 | 0.007 | 0.009 | 0.09 | 0.85 | 0.05 | 0.57 | 0.039 | 0.003 |
| 81A194 | 0.22 | 1.43 | 0.23 | 0.011 | 0.011 | 0.07 | 0.80 | 0.04 | 0.55 | 0.041 | 0.004 |
| 81A196 | 0.21 | 1.49 | 0.30 | 0.012 | 0.007 | 0.09 | 0.95 | 0.05 | 0.49 | 0.040 | 0.004 |
| 81A199 | 0.21 | 1.46 | 0.29 | 0.015 | 0.008 | 0.05 | 1.00 | 0.04 | 0.53 | 0.059 | 0.006 |
| 81A202 | 0.20 | 1.42 | 0.15 | 0.015 | 0.010 | 0.10 | 0.95 | 0.05 | 0.50 | 0.037 | 0.005 |

**Table 35.3.** Cross-sectional distribution of some elements in the ingot (%)

| Element* | Surface | One-half radius | Radius | Inhomogeneity |
|---|---|---|---|---|
| C | 0.206 | 0.210 | 0.200 | ±0.01 |
| S | 0.003 | 0.003 | 0.003 | 0 |
| P | 0.010 | 0.0096 | 0.0091 | ±0.0009 |

* Average of eight samples.

**Table 35.4.** Nonmetallic inclusions in ingots (relative units)

| Object tested | Oxides | Sulfides | Total concentration of inclusions (relative units) | Particle size |
|---|---|---|---|---|
| Standard | 2.5 | 2.5 | 4 | 5 |
| Ingot | 1 | 0.5 | 1.5 | 6–7 |

**Table 35.5.** Mechanical properties of the forging at 350°C

| Object tested | $\sigma_v$ (MPa) | $\sigma_T$ (MPa) | $\delta_i$ (%) | $\psi$ (%) |
|---|---|---|---|---|
| Standard | 560 | 350 | 18 | 50 |
| Forging | 630 | 475 | 26.8 | 75.0 |

**Table 35.6.** Impact resistance of the forging

| Object tested | $KCV^{-10}$ (J/m²) | EUSCV (J/m²) |
|---|---|---|
| Standard | 52 | 130 |
| Forging | 101; 175; 111; 110 | 247; 241; 250 |

Note: Test results are for four forged samples.

# 36 The Development of Electroslag Technology in Bulgaria

Ts.V. Rashev

Electroslag remelting (ESR), which was originally developed at the E.O. Paton Electric Welding Institute, is widely used throughout the world for the production of high-quality carbon and alloy steels and castings.

A commercial R-951 ESR installation that was designed at the E.O. Paton Institute and built at the Kadiev Machine Factory was put into operation in April 1975 in Bulgaria. Before this time, experiments were performed using a semi-commercial installation (designed and built in Bulgaria), which started operation in November 1972. These investigations resulted in the develoment of a technology for producing consumable electrodes and fluxes using locally available material, and also ESR of tool steels for forging die and extrusion components. Thus, when the commercial installation was put into operation, a number of Bulgarian plants had already used electroslag (ES) steel.

With the help of the E.O. Paton Institute and the "Dneprospetsstal" plant, and applying the institute's own experiences, it was possible to introduce ESR production of various steel brands, such as 5HNM, H12MF, 3H2V8F, 5H3V3MFS, 5HV2SF, 6HV2SF, 4H5MFS, etc., in a short period of time.

In Bulgaria it was most expedient to use consumable electrodes fabricated from castings with a length of 1,500 mm obtained by up-hill casting. These castings are assembled into electrodes of the required length by electroslag welding (ESW).

In order to develop remelting parameters for each type of steel, thorough analyses were performed to assess the quality and structure of the ingot, the type of nonmetallic inclusions, and their distribution and composition, and the oxygen concentration. ESR reduced the concentration of nonmetallic inclusions from 0.20 to 0.25 vol % in the electrode to 0.12 to 0.025 vol % (as measured by electrolytic precipitation). The anisotropy in longitudinal and transverse samples from ESR ingots was 0.9 to 1.0. The wear resistance of tools manufactured from cast ES alloy was 10% to 30% higher than that of tools made from imported forgings.

Since there was no suitable forging equipment in Bulgaria until 1984, some grades of steel were processed on the horizontal press. For example, guillotine cutters made of 5HV2SF and 6HV2SF steels were extruded at the Blagoi Popov Plant.

The experience obtained working on the R-951 installation showed the possi-

---

Author's affiliation: The Institute of Metallography and Technology, Academy of Science of Bulgaria, Bulgaria.

bility of further increasing production of ESR steel and of obtaining high-quality steel. The smooth surface of the ES castings permitted extrusion without preliminary roughing and conditioning. As a result, it was proposed to establish an ESR shop with a yearly capacity of 10,000 tons, which would be increased in the future.

The first part of the shop, consisting of four ESR-2.5VG type furnaces, was put into operation in 1983. The furnaces are equipped with both round and and square freezer molds with the following dimensions: 200, 250, 300, 350, 400, and 425 mm. The ESR ingots are subsequently hot worked in an SHL-55 radial forging machine built in 1984 by the Austrian company GFM. At present, the furnace control panel is being modified and the shop is being outfitted with computers and other electronic equipment.

A process for electroslag remelting under pressure (ESRP), which was developed by the ITM ASB, is now being introduced in Bulgaria for the remelting of nitrogen-alloyed steel.

A pilot ESRP installation working at a pressure of up to 50 atm in a nitrogen atmosphere started operations at the ITM ASB in 1976. This installation was used to test hermetic seals and assemblies impervious to gas. In 1984 the commercial R-951 installation was modified for operations at a pressure of 63 atm. The experience gained during two years' operation of this furnace led to the development of seals for ESR-2.5VG furnaces.

Several methods of measuring and controlling the ESRP process were developed by the use of such effects as the arc discharge that occurs in the slag layer during melting and the constant component of the melting current (rectification effect). High-precision sensors for measuring these effects were developed, which led to a better understanding of the ESRP process. In addition, these sensors provide feedback to various control devices that transmit precise and immediate control system reactions to random perturbations. This makes possible the complete automation of the entire process. (initiation–remelting–termination), which is important from the point of view of safety and performance of both the ESRP and the ESR processes.

The ESRP furnaces use steel consumable electrodes, which are produced from a nitrogen-bearing alloy (2.5% N) by casting under gas (this technique was developed by Bulgarian Academician A. Balevski and Corresponding Member I. Dimov). This type of alloy ought to be very pure and have a low segregation and a dense, homogeneous structure. Our investigations proved that it is possible to produce this type of alloy.

# 37 The Development of Electroslag Remelting at the "Huta Batory" Steel Mill from 1960 to 1985

M. Krutsinski, A. Novak, and V. Strama

## Construction of the ESR Installation at "Huta Batory"

The "Huta Batory" steel mill was the first in Poland to use electroslag remelted (ESR) steel. A pilot production installation was designed in 1959 and built and put into operation in 1960. The furnace was equipped with a single-phase transformer with a capacity of 100 KV-A that provided currents up to 2000 A and voltage regulation in the range of 40 to 70 V. Electrode feeding was performed manually. The water-cooled, copper freezer mold had an internal diameter of 125 mm and a length of 800 mm. The electrodes were 2,400 mm long and had diameters of up to 70 mm. The ingot was 700 mm long and weighed about 70 kg. The installation was equipped with a gas exhaust system.

This installation was used to study the effect of ESR on steel quality and to determine optimum remelting and steel macrostructure conditions as a function of the following remelting parameters: the current and the voltage, the electrode melting rate, the ratio of freezer mold diameter to electrode diameter, the slag chemical composition and volume, and the method of slag introduction.

In addition, the consumption of electric power and cooling water was investigated, along with installation productivity. Another study dealt with developing an exothermic slag composition for use in a commercial ESR installation. In particular, the following optimum parameters were determined: charge layer depth, 25 to 33 mm; exothermic slag composition, 37% slag, 33% steel chips, and 30% technical grade iron powder with a grain size of 0.06–0.15 mm; the best results were obtained using a slag containing 60% $CaF_2$ and 40% $Al_2O_3$. The slag should be dried before use at a temperature no lower than 300 to 400 °C. The current density and voltage were selected according to the diameter of the remelted electrode and the type of steel; the product of current and voltage was held constant. With appropriate parameters, the average power consumption is 1,400 KW-h/ton and the average cooling water consumption is about 70 $m^3$/ton of ingot.

The main problems that needed to be addressed in the first stages of the investigation were as follows: determining the amount of the exothermic mixture; determining the chemical composition of the slag and the amount of slag to use and how to introduce it into the furnace; determining the optimal voltage and current at different stages of remelting; and optimizing the electrode melting rate. This program was completed in 1963. The results were compiled in a special handbook.

---

Author's affiliation: "Huta Batory," Poland.

A commercial installation powered by a single-phase, 300-kV-A transformer was designed and built on the basis of the results of the above investigations. This installation included a transformer that could vary the voltage varied in 5-V steps between 45 and 65 V; protective and control devices and an electromechanical electrode positioner; two freezer molds with inside/outside diameters of 180/220 and 160/190 mm and a length of 1,450 mm (the mold sides and bottom were made of copper sheet with a thickness of 15 and 20 mm, respectively); and square electrodes with cross-sectional dimensions of 90 × 90 and 120 × 120 mm that were remelted in the smaller and larger mold, respectively. The electrode feed rate could be regulated between 10.5 and 19.5 mm/min; the cooling water pressure was from 0.20 to 0.25 MPa, and the ingots weighed 200 and 400 kg.

This installation was used to remelt bearing and high-speed steels to provide nitrogen and carbon enrichment to improve the physical properties of the steel. ESR reduced the size of all types of oxide and sulfide inclusions; eliminated hairline cracks in constructional alloy steels and local segregation in steels intended for nitrogen enrichment; and reduced carbide segregation in high-speed steels. The economic effect and the high quality of ESR steel, as well as growing need for ESR steel, made it necessary to increase production capacities and to widen the range of remelted steels.

For that purpose, construction of a new ESR shop was started at the mill. Construction was completed in 1970. The equipment consisted of three Soviet R-951 type installations, along with the installation described above, which was used up to 1976. This installation was mostly used for remelting high-speed and high-alloy steels (heat-resistant and acid-resistant steels). At present, there are about 130 ESR steels, and production is about 5,000 tons. The total spoilage is less than 0.5%, of which 40% is due to nonmetallic inclusions, 20% to cracking, 12% to internal porosity, 20% to casting discontinuities caused by interruptions of melting, and 8% to miscellaneous causes.

## Steel Quality and Properties

Rolled or forged electrodes are remelted using granular and lump slag with a grain size of up to 20 mm, although 80% of the grains are less than 0.5 mm in diameter. The concentration of $SiO_2$ (silica) in slags used for remelting steels containing aluminum, titanium, or other readily oxidizable elements was no greater than 1.8%. These steels were remelted under a protective argon atmosphere. The slag consumption was 30 to 40 kg/ton of remelted steel. For initiation of remelting, an exothermic mixture containing 60% working slag and 40% aluminum powder was used. Interruptions of remelting lasted no longer than 2 min. After remelting, ingots were soaked in the freezer mold for varying lengths of time, depending on their weight: 205-mm diameter ingots, 12 to 20 min; 320 × 320-mm ingots, 30 to 35 min; and 440 × 440-mm ingots, 40 to 45 min.

The roughly 13 types of remelted steels can be divided into the following groups: low- and medium-alloyed constructional steels for nitrogen and carbon enrichment and thermal hardening; bearing steels; stainless, acid-resistant, and heat-resistant steels; and high-speed tool steels. Separate, optimized remelting conditions were developed for each group: electrical parameters, slag, chemical composition of consumable electrodes, the type of argon atmosphere, melting rate, freezer mold cooling, etc. The effectiveness of ESR was determined by analyzing the ingot chemical composition, macro- and microstructure, mechanical characteristics, ductility, ability to harden, and corrosion resistance. Samples for analysis were taken from rolled ESR ingots after cropping 7% from the bottom end and 5% from the top end. Samples of the starting material were taken from electrode stubs that had undergone the same initial mechanical working. The results of these analyses have been published [1–15].

Similarly, samples for chemical analysis were taken from the initial material and from the top and bottom of the rolled ESR ingot after cropping. Samples were also selected from other parts of the ingot, as required.

A comparison of the chemical composition of the steel before and after remelting shows variations only in the concentration of such reactive elements as silicon, titanium, and aluminum. The concentration of other elements is within the error limits of the chemical analysis. A reduced silicon concentration is observed only in the case of steels that do not contain reactive elements (aluminum or titanium). The ESR of steels with a stabilized austenitic structure is characterized by relatively high losses of aluminum or titanium. For that reason, in order to retain these elements in the remelted alloy, their concentration in the consumable electrode should be increased or the slag should be deoxidized by adding aluminum powder during remelting. In these cases, it is preferable to remelt under argon and also to deoxidize the slag. The same methods should be used for titanium-stabilized, acid- and heat-resistant steels to reduce titanium loss. Nitrogen concentration in most of the remelted steels remains constant. Only in the cases of steels subjected to nitrogen enrichment, for example, 38HMYu, was a 30% reduction of nitrogen concentration observed; this was probably caused by an interaction between the nitrides and the slag.

Samples for the estimation of the nonmetallic inclusion content were cut from templets for macroetching. Estimates of nonmetallic inclusions in bearing steels were made according to Polish standard PN-H/84041, and according to PN-H/04510 for other types of steels. The results, given in Table 37.1, show that the size of all types of brittle inclusions after remelting was no greater than 1.0 points (irrespective of their size in the initial material) or that this type of inclusion was not detected. Sulfide and silicate inclusions show the same reduction in size. The size of elongated oxide inclusions in remelted steel is 0.5 points, and the size of spheroids is usually reduced by one-half. In bearing steels, the number of inclusions is reduced by 25%, and thus, the total oxygen concentration in the steel is almost unchanged.

The change in morphology of nonmetallic inclusions during ESR has a favorable effect on crack susceptibility. Magnetic particule inspection of samples of the initial and ESR alloy was performed. From 1 to 13 hairline cracks 2 to 4 mm long were detected in the initial alloy; however, no hairline cracks were observed in the remelted steel. Hairline cracks, which are discontinuities in the material, develop due to the presence of large nonmetallic inclusions, and can cause fatigue fracture and cracking and spalling if they intersect with the surface, which reduces contact fatigue resistance.

**Table 37.1.** Type and size of nonmetallic inclusions

| Steel brand | Type and size of nonmetallic inclusions | | | | | |
| --- | --- | --- | --- | --- | --- | --- |
| | Brittle | | Ductile | | Elongated oxides | Spheroids |
| | EO | BS | DS | S | PO | NDS |
| ShH15 | 1.5 (1.0) | 3.0 (0.5) | 0.5 (0.5) | 2.5 (0.5) | — | 4.0 (2.0) |
| ShH15SG | 3.0 (0.5) | 3.0 (—) | — | 2.0 (—) | — | 2.5 (1.5) |
| 12H2N4A | 3.0 (0.5) | 1.5 (0.5) | 1.0 (—) | 1.5 (0.5) | 1.0 (0.5) | 2.0 (1.0) |
| 18H2N4MA | 2.5 (0.5) | 1.0 (0.5) | — | 3.0 (1.5) | 0.5 (0.5) | 2.5 (1.0) |
| 20H2N4A | 3.0 (0.5) | 3.0 (—) | — | 2.5 (0.5) | 0.5 (0.5) | 2.5 (1.0) |
| 40HNM | 2.0 (0.5) | 2.5 (—) | — | 2.0 (1.0) | 0.5 (0.5) | 2.5 (1.0) |
| 38HMYu | 3.0 (0.5) | 3.5 (1.0) | — | 1.0 (0.5) | 2.5 (0.5) | 2.0 (2.0) |
| 2H13 | 1.5 (—) | 4.0 (—) | — | — | 3.0 (0.5) | 1.0 (1.0) |
| H17 | 2.5 (1.0) | 3.0 (0.5) | — | 1.5 (0.5) | 1.5 (—) | 1.0 (0.5) |
| 1H18N9T | 2.0 (—) | 3.0 (1.0) | — | — | 0.5 (0.5) | 1.5 (0.5) |
| SV7M | 3.0 (—) | 4.0 (—) | — | 1.0 (—) | 2.5 (0.5) | 2.5 (2.0) |
| | 2.0 (0.5) | 3.0 (—) | — | — | 0.5 (0.5) | 3.0 (1.5) |

Note: EO, elongated oxides; BS, brittle silicates; DS, ductile silicates; S, sulfides; PO, point-like oxides; NDS, nondeformable silicates. First numbers, before ESR; numbers in parentheses, after ESR.

Segregation and carbide banding in high-speed steel were estimated according to the test procedure in standard PN-H/93012, and according to PN-H/84041 for bearing steels. The samples were etched in nital (10% solution of $HNO_3$ in ethanol). Results are given in Table 37.2. It was determined that, due to specific solidification conditions during ESR, a high homogeneity is obtained throughout the ingot. Carbide agglomerations that are typical of hypereutectic Lederburite steels, which are usually connected through cavities, were not detected.

It is difficult or even impossible to reduce the size of carbide agglomerations by mechanical working of the steel, even with a large degree of deformation. This is shown by the presence of clearly defined carbide segregation bands in SV7M steel. After ESR, carbides have a size of 2 points and are uniformly distributed across the sample section. ESR improves the carbide distribution

**Table 37.2.** Estimates of segregation and carbide banding in high-speed steel

| Steel brand | Sample | Segregation | Banding | Standard |
| --- | --- | --- | --- | --- |
| SV7M | Initial ESR | 4.5 | — | PN-H/93012 |
| | Ingot bottom | 2.0 | — | |
| | Ingot top | 2.0 | — | |
| ShH15 | Initial ESR | 4.0 | — | PN-H/84041 |
| | Ingot bottom | 1.0 | — | |
| | Ingot top | 1.0 | — | |
| ShH15SG | Initial ESR | 1.0 | 2.5A | PN-H/84041 |
| | Ingot bottom | 0.5 | 0.5 | |
| | Ingot top | 0.5 | 0.5 | |

**Table 37.3.** Anisotropy coefficients of steel with and without remelting

| Steel brand | Anisotropy coefficients | | | |
| --- | --- | --- | --- | --- |
| | Tensile strength | Elongation | Reduction in area | Impact resistance |
| 12H2N4A | 1.0 (1.0) | 0.72 (0.89) | 0.63 (0.86) | 0.44 (0.71) |
| 18H2N4MA | — (0.99) | — (0.87) | — (0.85) | — (0.62) |
| 20H2N4A | 0.99 (1.0) | 0.75 (0.90) | 0.54 (0.81) | 0.49 (0.73) |
| 20HNM | 1.0 (0.98) | 0.73 (0.92) | 0.61 (0.85) | 0.59 (0.75) |
| 38HMYu | 1.0 (0.99) | 0.87 (0.94) | 0.86 (0.88) | 0.78 (0.82) |
| H17 | 0.93 (0.98) | 0.77 (0.85) | 0.92 (0.95) | 0.64 (0.65) |

Note: First numbers, initial alloy; numbers in parentheses, ESR alloy.

in bearing steels, although the number of primary carbides in these steels is significantly reduced.

The mechanical properties of remelted steels were tested in longitudinal and transverse samples. Transverse samples were taken at a distance of one-quarter of the diagonal of the product. Tensile test samples had a diameter of 20 mm and impact test samples were square, with dimensions $20 \times 20$ mm. Heat treatment of samples was in accordance with standard practice for this type of steel. The increase in steel purity improved the mechanical properties, especially the tensile strength and ductility of samples transverse to the direction of plastic deformation. This is evident from the anisotropy coefficients shown in Table 37.3.

For H17 steel, additional testing of corrosion resistance was performed according to procedure PN-H/04630. It was determined that ESR does not affect the corrosion resistance of this type of steel, since corrosion is mainly determined by the chemical composition, which is not changed by remelting.

## Conclusions

ESR significantly improves steel quality. The ESR steel is characterized by a homogeneous chemical composition and a dense structure. In addition, it does not have macrostructural defects, such as axial porosity, shrinkage pipes, and segregation cones. The remelting process is accompanied by a loss of such elements as silicon, titanium, and aluminum. The concentration of these elements can be regulated by deoxidizing the slag during ESR. The concentration of titanium and aluminum in the steel can also be retained by increasing their concentrations in the consumable electrode. The loss of these elements can be reduced by using a protective argon atmosphere.

The desirable service characteristics of ESR steel, which are due to its purity, make the process economically attractive for a wide range of applications. At present, preparatory work is being carried out at the "Huta Batory" steel mill in order to increase electroslag steel production.

# 38 The Experience of the "Huta Beldon" Steel Mill in Electroslag Remelting

## H. Shvei and S. Mista

## Introduction

For a long period of time, the "Huta Beldon" mill has specialized in the production of alloy steels used to manufacture high-reliability machine parts and equipment for use in the automobile, chemical, machine tool, ship building, and power industries, and other industries. In order to satisfy the continually increasing needs for high-quality steel, it was found to be necessary to improve production methods and to develop new ones.

For this purpose, an R-951 electroslag remelting (ESR) installation with a yearly ingot production capacity of 1,500 tons was purchased in 1969 from the Soviet Union and was installed at "Huta Beldon" with the help of Soviet engineers.

The good results obtained with the R-951 installation and the improved steel quality led to the purchase of two more ESR installations in 1978: a Soviet ESR2.5VG furnace, with a capacity of 1,500 tons/y, and an 8Mg installation built by the French company "Ertei." At present, ESR accounts for 7% of the total steel production at "Huta Beldon." This is 55% of the total production of ESR steel in Poland.

## Characteristics of the ESR Installations

### Installations R-951 and ESR-2.5VG with Stationary Freezer Molds

These are single-column installations with a stationary water-cooled, copper freezer mold. The R-951 installation is equipped with a single-phase transformer with a capacity of 1 MV-A. The primary voltage is 6,000 V, the working voltage is 41–95 V, at a current of 14,000 A. The voltage is regulated in 17 steps. The installation is used to produce ESR ingots with cross-sectional dimensions of $320 \times 320$ mm and a weight of 1 ton.

The ESR 2.5VG installation is also equipped with a single-phase transformer with a capacity of 1.6 MV-A. The primary voltage is 6,000 V, the working voltage is 31 to 116 V, and the maximum current is 17,000 A. The voltage is regulated in 49 steps. The installation is used to produce 1- and 2-ton ingots with dimensions of $320 \times 320$ and $400 \times 400$ mm, respectively.

Most of the electroslag (ES) ingots are hot or cold rolled.

---

Authors' affiliation: "Huta Beldon" Steel Mill, Poland.

## The 8Mg Installation with a Movable Freezer Mold

This is a two-column installation with a movable freezer mold that is attached to one of the columns. Carriers for the electrode holders are fixed to each column. The electrode holders are equipped with hydraulic clamps. The carriers position the electrode vertically and the electrode holders can move in a horizontal arc. Electrodes with welded-on, reusable stubs are used so that several electrodes can be remelted per ES ingot. The installation is equipped with two single-phase transformers with a rating of 1.8 MV-A; the first is a step-down from 6,000 V to 850 V, and the second transforms the 850-V input to the 50- to 120-V operating voltage at a current of up to 14,000 A. The transformers are connected in series. Voltage regulation is continuous.

The installation uses three types of freezer molds to produce ES ingots up to 4,500 mm in length: a 600-mm diameter mold for ingots weighing up to 10 tons, a 350-mm mold for ingots weighing up to 4.5 tons, and a 300-mm mold for ingots weighing up to 2.5 tons.

This two-column installation with a movable mold permits adjustments of ingot weight in accordance with the weight of the final product, or of the length to give a whole number of pieces if the ingot is to be cut apart to provide a high product yield.

The copper freezer molds are tapered toward the bottom with a taper of 0.7% (one side) and a length of 700 mm. The mold and base plate are cooled by a closed-cycle water cooling system.

## Slags Used for ESR

At first, slags ANF-6 and ANF-1P were used for ESR. These slags were characterized by a high specific power consumption and a low desulfurization ability. In order to overcome the limited availability of these slags and to increase the efficiency of ESR, the 2022-type slag, with an increased amount of CaO, was tested. Favorable test results led to the adoption of this new slag for steel production. When the new installations were put into operation in 1978, the ESR process for the 8Mg installation with a movable freezer mold used the type 2015 slag.

Tests of the 2015 slag showed that specific power consumption, desulfurization, and removal of nonmetallic inclusions were good, and so the slag was adopted for ESR steel production. The chemical composition of the slags used for ESR is given in Table 38.1.

The use of the 2015 slag can enrich hydrogen concentration in the steel during remelting, especially in the bottom end of the ingot. In order to limit the detrimental effects of hydrogenation and to reduce the concentration of gases and impurities with a high vapor pressure, flaking-susceptible steels are vacuum degassed in the VAD installation.

**Table 38.1.** Chemical composition of slags used for electroslag remelting (%)

| Slag type | CaF$_2$ | Al$_2$O$_3$ | CaO (+MgO) | MgO (max) | SiO$_2$ (max) | FeO (max) | S (max) |
|-----------|---------|-------------|------------|-----------|---------------|-----------|---------|
| ANF-6     | 67      | 27          | 3          | 1         | 2             | 0.5       | 0.03    |
| S 2022    | 59      | 23          | 15         | 3         | 2             | 0.5       | 0.03    |
| S 2015    | 33      | 32          | 31         | 3         | 2             | 0.5       | 0.03    |

## Slag Melting

ESR is initiated with a solid slag charge. Until 1979 a short-circuit method with an exothermic mixture of steel turnings was used. This method yields a very low slag melting rate (40 kg of slag in 20 to 30 min). There were instances when the current was interrupted because solid slag entered the gap between the electrode and the steel charge. These drawbacks were eliminated in 1978 by arc melting the slag. In arc melting, a so-called steel starter is introduced into the gap between the electrode and the charge. Slag portions weighing from 1 to 3 kg are introduced at the moment of arc initiation, and more portions are added as slag melting proceeds. Arc melting is then followed by ES remelting. The time for slag melting is less than 7 to 12 min. This method yields a smooth, better refined ingot bottom and reduces power consumption.

## Remelting and Its Termination

Most of the process, that is, steady-state remelting, is performed at a uniform ingot solidification rate, which is determined according to the equation $R = 0.98 \times A$ (kg/h) for square freezer molds, where $A$ is the width of freezer mold in mm, and $R = 0.85 \times D$ (kg/h) for circular freezer molds, where $D$ is the diameter of the freezer mold in millimeters.

The voltage, current, and power are selected to satisfy these requirements. As remelting proceeds and the ingot length increases, the electrical resistance of the system also increases, and the chemical composition of the slag changes. Thus, the drop in voltage continues to increase, and the current decreases. These electrical parameters are regulated manually.

The termination of the process is of great importance in preventing the formation of shrinkage pipes. Initially, a smooth process termination was used in which the power was periodically reduced every 3 to 5 min. The main drawback of this method was the formation of a shrinkage pipe, due to the difficulty of controlling the electrical parameters manually. Since 1978 a three- to four-step reduction of melting power has been used. During the first step, the power is reduced from 60% to 70%; during the next step, from 9% to 11%; and during the last step, from 5% to 10%. This method provides a riser 30 to 60 mm high and prevents the formation of shrinkage pipe, which reduces the amount of material that must be cropped from the ingot top.

After termination of remelting, the ES ingots are slowly cooled. The cooling method depends on steel type and ingot weight.

## Special Problems in ESR

In parallel with the development of ESR steel production, other studies to improve remelting techniques and alloy structure and purity and to reduce specific power consumption and product cost were conducted.

### The Effect of Chemical Composition of the Slag on the Purity of ES Steel

Tests were performed with 40HNMAZ steel. Remelting was performed in a stepped, water-cooled, copper freezer mold with a cross section of $300 \times 300$ mm and a length of 1,300 mm. All the steel for ESR was produced in an electric arc furnace and was subsequently vacuum degassed and cast into square ingots weighing 980 kg. The ingots were rolled into electrodes with dimensions of $165 \times 165$ or $200 \times 200$ mm, depending on the type of remelting.

After annealing and etching, the electrodes were remelted using different amounts of slag of various compositions. The ANF-6, S2022, and S2015 slags were used (Table 38.1).

The slag was melted and ground to a grain size of from 1 to 30 mm. During the initial melting, the slag was deoxidized with granular aluminum (0.2 kg/melting run); it was not deoxidized during remelting. The melting parameters were selected to provide a remelting rate of 280 to 300 kg/h. The characteristics of some ESR runs are given in Table 38.2. These include the quantity of slag and its composition, the dimensions of the electrode the power required for remelting, and the specific power consumption.

To determine the micro- and macroinclusion content of remelted alloy, the electrode alloy and the remelted alloy were analyzed. Samples were taken at four-fifths of the electrode height and from the top end of the rolled ingots, which measured $120 \times 120$ mm.

The nonmetallic inclusion content was estimated by macroetching in 50% HCl, according to procedure PN-57/H-04501. No defects were observed in any remelted samples, irrespective of the ESR run.

**Table 38.2.** Parameters of various electroslag remelting processes

| ESR | Slag type | Slag consumption (kg/ton) | Electrode transverse cross section (mm) | Filling factor, freezer mold | Power consumption (KW-h/t) | Remelting rate (kg/h) |
|-----|-----------|---------------------------|------------------------------------------|------------------------------|----------------------------|------------------------|
| 1 | ANF-6 | 45 | $165 \times 165$ | 27 | 1,683 | 280 |
| 2 | S 2022 | 45 | $165 \times 165$ | 27 | 1,575 | 280 |
| 3 | S 2022 | 35 | $200 \times 200$ | 39 | 1,381 | 290 |
| 4 | S 2015 | 35 | $200 \times 200$ | 39 | 1,204 | 295 |

**Table 38.3.** Averages of the maximum size of nonmetallic inclusions in 40HNMA steel according to standard PN-71/H-04510

| Remelting process | Type and size of nonmetallic inclusions | | | | | | | |
|---|---|---|---|---|---|---|---|---|
| | EO | PO | DS | BS | NDS | S | PN | AN |
| 1 | 3.5 (1.5) | 0 (0) | 0 (0) | 2 (1) | 1 (1) | 3 (1) | 0 (0) | 0 (0) |
| 2 | 3.5 (1.5) | 0 (1) | 0 (0) | 2 (1) | 1 (0) | 3 (1) | 0 (0) | 0 (0) |
| 3 | 4 (1.5) | 1 (1) | 0 (0) | 2 (1) | 2 (1) | 3.5 (1) | 0 (0) | 0 (0) |
| 4 | 3 (1) | 1 (1) | 0 (0) | 1.5 (1) | 1.5 (0) | 2 (1) | 0 (0) | 0 (0) |

Note: EO, elongated oxides; PO, point-like oxides; DS, deformable silicates; BS, brittle silicates; NDS, nondeformable silicates; S, sulfides; PN, point-like nitrides; AN, aluminum nitrides. Numbers, starting steel (electrode); numbers in parentheses, steel from the ingot top.

The nonmetallic inclusion content was also estimated according to the scales of standard PN-57/H-04501. For this purpose, microsections were cut from transverse samples taken along the ingot radius, from the surface to the center. The results are given in Table 38.3.

It is clear from these results that ESR provides more highly refined steel and reduces its content of elongated oxides, brittle silicates, and sulfides. The use of slags containing from 15% to 32% CaO gives the same alloy purity as a $CaF_2$–$Al_2O_3$ slag and increases the degree of desulfurization by 80%.

Similar results were also obtained for the gas content in the steel, with the exception of hydrogen, which increased in concentration from 0.5 to 1.0 ppm, especially in the bottom end of the ingot. The mechanical properties, including anistropy, were also improved by remelting.

## Power Consumption

The data in Table 38.2 demonstrate the significant effect that slag chemical composition and the freezer mold filling factor have on specific power consumption in ESR. ESR using ANF-6 slag had the highest specific power consumption. Replacing ANF-6 by S-2022 slag reduces power consumption by 108 kW-h/t while the melting rate (280 kg/h) and the filling factor (0.27) remain the same.

The use of larger electrodes that increase the filling factor to 0.39 further reduced power consumption to 194 kW-h/t, and reduced slag consumption by 10 kg/t, that is, 22.6%. In addition, remelting productivity increased slightly.

The use of S2015 slag instead of S2022, with a filling factor of 0.39, reduced power consumption by a further 177 kW-h/ton, that is, by 12.8%, with an insignificant increase in productivity.

Such a significant reduction in power consumption was obtained due to the high resistivity of slags containing CaO. The increased generation of heat in the slag reduces electric power consumption during ESR.

H. Shvei and S. Mista

**Table 38.4.** Change in slag chemical composition during electroslag remelting with and without slag deoxidation

| Series | Remelting run | Sample from ingot | Slag chemical composition | | | | | | | | Percent sulfur contents in electroslag steel | Percent sulfur contents in electrodes |
|---|---|---|---|---|---|---|---|---|---|---|---|---|
| | | | CaO | CaF$_2$ | Al$_2$O$_3$ | MgO | SiO$_2$ | FeO | Cr$_2$O$_3$ | S | | |
| I | | | *Without slag deoxidation* | | | | | | | | | |
| | 1 | Top | 33.12 | 30.14 | 30.98 | 2.57 | 2.56 | 0.50 | 0.10 | 0.027 | 0.007 | 0.016 |
| | | Middle | 31.66 | 24.28 | 27.44 | 2.39 | 10.90 | 1.80 | 0.90 | 0.010 | 0.010 | 0.010 |
| | | Bottom | 28.93 | 20.92 | 23.88 | 2.33 | 18.94 | 3.30 | 2.12 | 0.012 | 0.015 | |
| | 2 | Top | 31.05 | 34.67 | 28.15 | 2.82 | 2.38 | 0.75 | 0.15 | 0.025 | 0.007 | 0.014 |
| | | Middle | 29.17 | 30.22 | 27.58 | 2.36 | 9.10 | 1.30 | 0.38 | 0.021 | 0.009 | |
| | | Bottom | 28.12 | 25.99 | 25.75 | 2.10 | 15.56 | 1.69 | 0.78 | 0.013 | 0.016 | |
| | 3 | Top | 27.42 | 33.02 | 34.76 | 2.21 | 1.80 | 0.69 | 0.07 | 0.028 | 0.005 | 0.014 |
| | | Middle | 24.96 | 26.29 | 32.18 | 2.17 | 12.56 | 1.12 | 0.70 | 0.021 | 0.009 | |
| | | Bottom | 24.32 | 22.92 | 31.16 | 2.10 | 16.70 | 1.67 | 1.12 | 0.014 | 0.012 | |
| II | | | *With slag deoxidation* | | | | | | | | | |
| | 1 | Top | 30.60 | 34.86 | 30.43 | 1.08 | 2.42 | 0.52 | 0.17 | 0.020 | 0.005 | 0.021 |
| | | Middle | 30.23 | 30.86 | 34.10 | 1.04 | 2.83 | 0.77 | 0.15 | 0.019 | 0.004 | |
| | | Bottom | 30.96 | 27.88 | 37.80 | 0.98 | 1.61 | 0.56 | 0.18 | 0.026 | 0.004 | |
| | 2 | Top | 28.92 | 36.26 | 29.18 | 2.97 | 1.86 | 0.56 | 0.22 | 0.028 | 0.005 | 0.023 |
| | | Middle | 28.60 | 33.36 | 32.45 | 3.07 | 1.56 | 0.66 | 0.28 | 0.018 | 0.004 | |
| | | Bottom | 27.85 | 29.33 | 36.97 | 3.12 | 1.70 | 0.62 | 0.29 | 0.021 | 0.005 | |
| | 3 | Top | 31.16 | 31.59 | 32.34 | 2.36 | 1.60 | 0.78 | 0.15 | 0.025 | 0.004 | 0.009 |
| | | Middle | 31.82 | 28.40 | 34.94 | 2.44 | 1.50 | 0.69 | 0.19 | 0.026 | 0.004 | |
| | | Bottom | 32.34 | 25.62 | 36.79 | 2.38 | 1.80 | 0.80 | 0.25 | 0.019 | 0.004 | |

## Changes in Slag Chemical Composition in Remelting Long Ingots

The change in slag chemical composition during ESR was analyzed under commercial production conditions: Remelting electrodes with a diameter of 320 mm and a weight of 900 kg were used to produce ingots with a diameter of 600 mm, a length of 4,500 mm, and a weight of up to 10 tons.

WCL (4H5MFS) steel was remelted in the 8Mg installation using slag S2015 (83 kg/melting run) and annealed, unconditioned electrodes.

In the course of the study, several remelting series with three runs in each series were performed. In the first ESR series, no slag deoxidation was performed; in the second series, the slag was deoxidized with granular aluminum that was added at 30-s intervals (1 kg of aluminum/ton of alloy)by an injector with an automatic scale.

Samples of slag and steel were taken for chemical analysis after melting of the slag and the remelting of about 600 kg of steel (test A), after one-half of the ingot had solidified (test B), and just prior to termination of remelting (test C). Test results are given in Table 38.4.

In remelting annealed, unconditioned electrodes without slag deoxidation, the $CaF_2$ concentration in the slag is reduced from 33% to 23%, the concentrations of CaO, MgO, and, to lesser extent, $Al_2O_3$ and S are also reduced.

The relatively slight change in $Al_2O_3$ concentration can be explained by an oxidation of the aluminum that is dissolved in the alloy, which then remains in the slag as $Al_2O_3$. The $SiO_2$ concentration in the slag increases the most, from 2% to 17%. The concentration of FeO (iron oxide concentration expressed as FeO) and $Cr_2O_3$ also increase.

When the same steel (WCL) is remelted with similar remelting parameters but in the presence of slag deoxidation (1 kg of aluminum/ton of alloy), the $CaF_2$ concentration increases from 24% to 25%, and the concentration of $Al_2O_3$ increases from 30% to 36%, but the concentration of the other components does not change significantly. In the absence of slag deoxidation, the weight of slag increases slightly because the increase in the concentrations of $SiO_2$, iron oxides, and other elements is not completely compensated for by evaporation of fluorides and by slag deposition on the freezer mold walls. However, when the slag is deoxidized, its weight decreases slightly.

## Desulfurization During Remelting of Long ES Ingots

ESR with S2015 slag ($CaF_2-CaO-Al_2O_3$) is characterized by a significant desulfurization of the steel (Table 38.4).

The data on steel desulfurization that occurs when long ESR ingots are remelted, both with slag deoxidation by aluminum and with no deoxidation, are presented in Table 38.5.

When S2015 slag, containing 2% to 4% $Sio_2$, is used, the sulfur concentration in the remelted steel is 0.004% to 0.005% and the degree of desulfurization is between 60% and 80%, depending on the initial concentration of sulfur. In-

**Table 38.5.** Dependence of the degree of steel desulfurization during ESR on slag deoxidation by aluminum (%)

| Slag | ESR run | Test sample location on ingot | | |
|------|---------|------|--------|--------|
|      |         | Top  | Middle | Bottom |
| Not  | 1       | 56.2 | 37.2   | 6.2    |
| deoxidized | 2 | 50.0 | 35.7   | —      |
|      | 3       | 64.3 | 35.7   | 14.3   |
| Deoxidized | 1 | 76.2 | 31.0   | 81.0   |
|      | 2       | 78.3 | 82.6   | 78.3   |
|      | 3       | 55.5 | 55.5   | 55.5   |

creasing the $SiO_2$ concentration in the slag to 9% to 12% results in an increased sulfur concentration of 0.008% to 0.012% and a drop in the degree of desulfurization from 35% to 40%.

No desulfurization occurs in ESR at a $SiO_2$ concentration from 16% to 19%. Remelting with slag S2015 gives a high degree of steel desulfurization, and this slag can be used for remelting long ingots, provided that the electrodes have undergone surface conditioning and that the slag is deoxidized during remelting.

## Conclusions

The introduction of ESR for steel production at the "Huta Beldon" steel mill led to significant improvements in the service properties of a wide range of steel types, thus satisfying the increasing needs for steel products. The use of ES installations with both movable and stationary freezer molds makes it possible to obtain ingots of the required weight. The installation with a movable freezer mold is primarily used to produce steels that are susceptible to hot cracking. Our studies, which were conducted under production conditions, have led to improvements in the reliability and efficiency of alloy steels.

# 39 The Development of Electroslag Remelting at the "Zhelezarna Ravne" Metallurgical Plant in Yugoslavia

J. Rodich and M. Schvaiger

"Zhelezarna Ravne" is one of three metallurgical plants belonging to the "Slovenski metallurgical plants" association. The first electroslag (ES) products in Yugoslavia were produced here in 1973. The electroslag remelting (ESR) shop is equipped with a Soviet R-951 ESR furnace for the production of 400-mm diameter ingots, weighing up to 2 tons and using consumable electrodues with a maximum length of 5 m. In order to increase the capabilities of the installation, it was modified in 1974 by the introduction of short, movable freezer molds with round and square cross sections from 220 and 500 mm. The initial production technology was also modified in order to reduce slag consumption and electric power consumption.

With the appearance of secondary metallurgical processing (VAD, VOD), ESR came to be regarded as a tertiary refining process that is oriented toward controlling the solidification and the isotropy of the steel to obtain specified properties.

A new installation, built by the "Inteco" company (Austria), was put into operation at the end of 1982. The main specifications of the furnace are as follows: a short, stationary freezer mold; a movable bottom plate, and the provision for electrode replacement during remelting; maximum ingot length 6 m; maximum ingot weight, 36 tons; ingot cross section, round with a diameter from 500 to 1,000 mm or rectangular with the dimensions 500 × 1,000 mm; current, 25 kA; power, 3250 kV-A; power line frequency, 50 Hz; and secondary voltage, 30 to 130 V (100 regulation steps).

This ESR furnace played an important role in the evolution of the ES installations (Fig. 39.1)—from a long, stationary mold followed by a short movable freezer mold with replaceable electrodes to a short, stationary freezer mold with the capacity for electrode replacement during remelting, a movable bottom plate, sliding current leads, and automatic electrode depth control. This last furnace, which is the latest development of the "Inteco" company, was first used at the "Zhelezarna Ravne" Plant. The furnace design has overcome the main drawbacks of previous models (Table 39.1). The parallel power supply busbars and sliding current contacts (Fig. 39.2) permit the use of remelting currents of over 50 KA at the power line frequency.

Significant advances in computer automation of ESR provide a high-quality

Authors' affiliations: J. Rodich, Slovenski metallurgical plants, Lyublyana, Yugoslavia; M. Schvaiger, "Zhelezarne Ravne" Metallurgical Plant, Ravne-na-Koroshke, Yugoslavia.

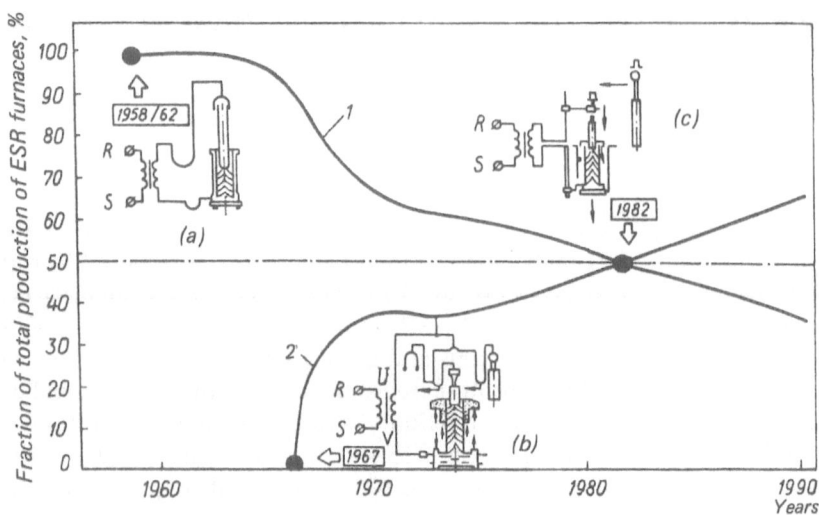

**Fig. 39.1.** Diagram of the fraction of total production for different furnaces (by year): (a) furnace with a stationary freezer mold (introduced in 1958 to 1962); (b) furnace with a short, movable freezer mold and electrode replacement (introduced in 1967); and (c) furnace with a short stationary freezer mold, electrode replacement, movable bottom plate, sliding contacts, and automatic control of the electrode penetration depth (introduced in 1982). (1) Long freezer molds; and (2) short freezer molds.

**Table 39.1.** Comparative characteristics of modern electroslag remelting equipment

| Characteristic | Type of installation | | | |
|---|---|---|---|---|
| | Stationary freezer mold | Coaxial or bifilar power supply and stationary freezer mold | Short, movable freezer mold and electrode change | Short, movable freezer mold, electrode change, and movable bottom plate |
| Ingot length | − | − | + | ++ |
| Electrode length | −− | −− | + | + |
| Electrode change | − | − | + | + |
| Repair of copper freezer molds | −− | −− | + | + |
| Access to the slag layer | − | −− | ++ | ++ |
| Additions to the slag | + | + | − | ++ |
| Furnace inductance | − | ++ | −− | + |
| Cost of freezer mold | − | − | + | + |
| Slag consumption | − | − | + | + |
| Yield of steel | − | − | + | ++ |
| Furnace productivity | − | − | + | + |
| Production intricate shapes (rolls) | − | − | + | + |

Note: ++, very good; +−, good; −, poor; −−, very poor.

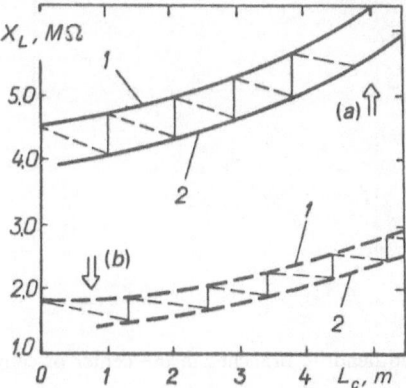

**Fig. 39.2.** Reactance of an electroslag remelting furnace with a movable freezer mold, an opposing motion of the electrode and mold, and flexible power cables (a); reactance of furnace with a movable bottom plate, busbars, and sliding contacts (b); curves 1 and 2, beginning and end of remelting, respectively; ingot diameter, 800 to 900 mm; current, 13 to 20 kA; frequency, 50 Hz.

surface and an optimal structure of the ingot due to continuous regulation of the electrode penetration depth and control of the ingot solidification rate.

Long ESR ingots (Fig. 39.3) are cut into semifinished products of optimal length for subsequent forging operations. The production of long ingots reduces slag consumption and increases the installation's productivity and product yield. Semifinished products that are cut from unforged, heat-treated ESR ingots are used directly for manufacturing tools and various components. At present, a third ESR furnace, which is smaller than the two previous models, is being built, and a fourth one is in the planning stages. New developments, such as remelting

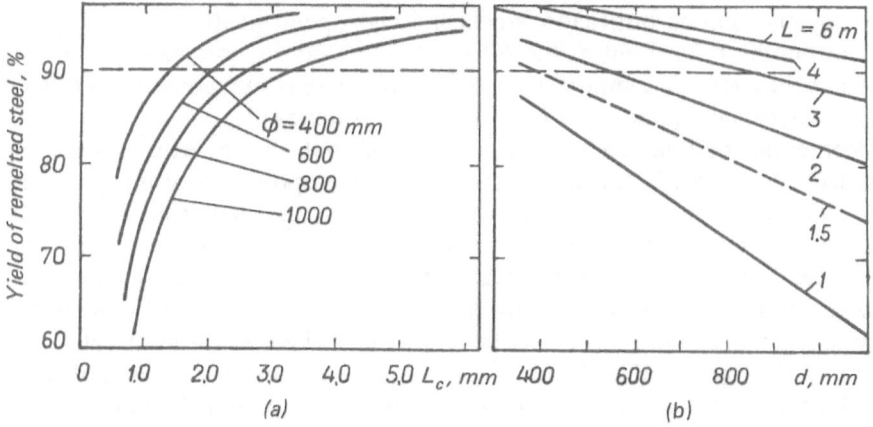

**Fig. 39.3.** Dependence of the yield of remelted steel on the ingot (a) diameter and (b) length.

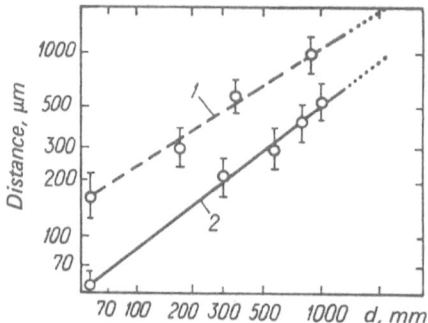

**Fig. 39.4.** The distance between dendrite branches in the center of conventionally produced (1) ingots and (2) ESR ingots.

in a protective atmosphere and adjusting the chemical composition of the slag and the steel during remelting, are being incorporated into the design.

Recently, the furnace design and control systems were improved, which allowed the commercial introduction of a group of specialized techniques that are expected to be of great importance.

If the degree of desulfurization, homogeneity of chemical composition and of temperature, and freedom from nonmetallic inclusions can be provided by secondary metallurgical processes, then control of the micro- and macrostructure of the ingot can be achieved just by proper control of the solidification process. Thus, remelting processes are expected to become more important in the production of high-quality materials when remelting is combined with other secondary metallurgical processes, due to the favorable effect of ESR on the mechanical and service characteristics of alloy.

For example, the structure of a 1-m diameter ESR ingot is equivalent to that of a conventionally produced ingot with a diameter of 300 mm, as measured by the distance between dendrite branches (Fig. 39.4). It is possible to obtain more homogeneous Lederburite tool steels and high-speed steels.

Continuing refinement of the technology, based on systematic studies led to the development of computer-based control systems for remelting that guarantee the necessary quality and reliability of production.

Remelting of long castings is economical, but special new techniques are needed. During ESR, oxidizable elements are transported from the steel into the slag, whereas other elements are incorporated into the steel from the slag, due to reducing reactions. Aluminum and silicon are the most reactive elements.

Under normal conditions, the reactivity of $SiO_2$ in the slag increases as remelting proceeds, whereas the concentration of aluminum and its reactivity decrease. The oxygen concentration in the bottom end of the ingot is no greater than 10 ppm; in the top end of an 6-m ingot, it is four times higher. The loss of silicon in the bottom end of an ingot is, on the average, 30% greater than in the top end.

When remelting 6-m long ingots (Fig. 39.5), the initial $SiO_2$ concentration (less than 2%) in the slag increases to 18% at the end of remelting. The slag

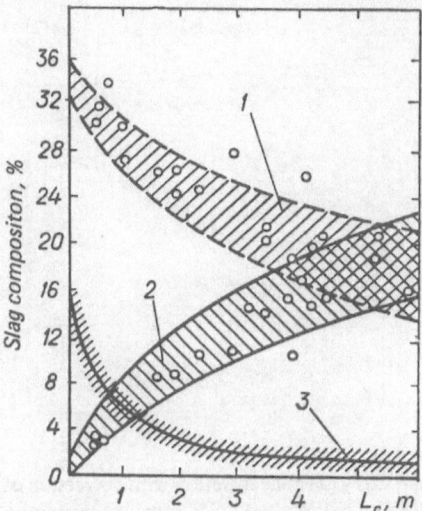

**Fig. 39.5.** Change in the slag composition when long ingots are remelted and when aluminum is used for slag deoxidation: (1) $CaF_2$; (2) $SiO_2$; and (3) basicity a $(CaO)/(SiO_2)$. Ingot diameter, 1 m; steel, Cr–Ni–Mo; initial slag composition, 30% $Al_2O_3$, 30% CaO to 33% $CaF_2$, and < 2% $SiO_2$; solidification rate, 900 kg/h.

basicity is sharply reduced from 16 to below 2. The $CaF_2$ concentration in the slag drops from 33% in the initial stages of remelting of 6-m long ingots to less than 20% at the end of the process. As casting proceeds, the melting point, viscosity, and resistivity of the slag increase due to changes in its chemical composition.

When remelting low melting point steels, the slag solidifies first, and slag particles can be entrapped in the molten steel, which results in the formation of large slag inclusions in the top end of the ingot. In order to provide successful termination of remelting, the slag melting point should be at least 100 °C lower than the melting point of the steel.

The reduction of slag basicity during ESR slows the matallurgical reactions, resulting in a continuous reduction of the desulfurization ability, from 80% to 90% at the bottom end of the ingot to 10% and less at the top end. As was mentioned earlier, the oxygen concentration in the ingot increases along its length, from 10 ppm or less in the bottom end to over 40 ppm in the top.

When remelting is performed without a protective atmosphere, hydrogen concentration increases; remelting in a protective atmosphere practically suppresses this increase.

In order to guarantee chemical homogeneity along the ingot length, special measures should be provided for maintaining a constant chemical composition of the slag during remelting. However, continuous correction of the slag composition in the course of remelting is rather difficult.

In order to reduce changes in the chemical composition along the length of the

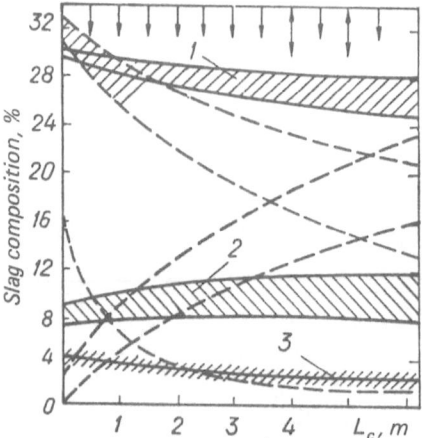

**Fig. 39.6.** Change of $SiO_2$ (1) and $CaF_2$ (2) concentration with correction of the slag composition by method III; basicity $b = (CaO)/(SiO_2)$ (3); dashed lines, previous technique.

ingot, special methods are used for remelting electrodes of different or identical chemical composition.

Variations in slag composition lead to changes in the chemical composition of the steel. Three methods were developed for correcting the slag composition during remelting; they differ in purpose and in the degree of interaction with the process.

Method I, which is the cheapest and simplest, is the most widely used. Method II is more advanced than I; however, in this article we describe only method III (Fig. 39.6), which gives the best results, although it is not the only one in commercial use. The third method is based on the fact that practically every ESR ingot is remelted under a highly basic slag only during the initial stages of the process, after which the slag basicity falls to 2. Therefore, at the beginning of the process, a portion of the used slag is introduced into the fresh slag, whereas later CaO and $CaF_2$ are added and a portion of the slag is replaced by fresh slag.

The advantages of the third method are as follows: The specific slag consumption is reduced, gas enrichment of the steel is also reduced, the higher concentration of $SiO_2$ in the slag reduces the loss of silicon in the bottom end of the ingot, and a lower melting current can be used.

This method allows the remelting of 6-mm long ingots under conditions of improved slag and steel homogeneity.

Remelting of electrodes by various methods is of great interest from an economic standpoint, but there are certain technical problems. Usually, it is insufficient to sort electrodes according to concentration of the most important elements; thus, additional alloying of the steel during ESR is often necessary. There are two main ways to introduce alloying elements into the freezer mold: (1) introduction of additives either by means of a continuous injector or manually, and (2) the use of an additional electrode with a higher concentration of the desired alloying elements.

For successful additional alloying, the following measures should be taken: The melting point of the ferroalloy must be very close to the steel melting point; the density of the ferroalloy must be equal to or lower than the steel density and higher than that of the molten slag; ferroalloys for introduction by injection should have a uniform grain size; and additives should be introduced either continuously or at very frequent intervals.

The degree of additional alloying is limited by the remelting conditions and depends on the diameter of the ingot. We have succeeded in increasing the concentration of Si by 0.30%, Cr by 0.60%, V by 0.20%, and C by 0.15%.

Our technology, modern installation design, and automatic control system were used as a basis for developing ESR further. The experience gained in remelting in a protective gas atmosphere above the slag layer should also prove useful.

A technology for producing hollow ingots is not yet completely developed. However, it is doubtful that this technology can be used economically for commercial production because of its limited potential.[1]

Specialists in Yugoslavia are now developing ESR for producing complex shaped rolls and machine tools that do not require forging.

---

[1] A technique for producing hollow ingots using a piercing method with sequential electrode feed has been developed in the Soviet Union [editor's note].

# 40 Electroslag Remelting of Heat-Resistant Alloys: Thermal Balance of Melting and Alloy Chemical Homogeneity

J. Domingue and K.O. Yu

## Introduction

The development of electroslag remelting technology (EST) at Special Metals Corporation (SMC) was started in 1968 when the the first laboratory installation built by the Consarc Corporation was installed. This installation had a capability for producing ingots with a maximum diameter of 200 mm by means of electroslag remelting (ESR) and vacuum arc remelting (VAR). The chemical and structural homogeneity, purity, deformability, and forging properties of ESR ingots with a diameter of up to 150 mm are similar to those of VAR ingots. At that time, there was no demand for ESR products in the aerospace industry; VAR was the only appropriate remelting process for producing alloys used for manufacturing rotating components of aircraft engines produced in the United States [1–3].

The commercial production of ESR alloy at SMC was started in 1976 [4] when a commercial ESR furnace was installed. This furnace replaced radiofrequency induction melting (RFIM) in the production of higher-quality tools and stainless steels and corrsion-resistant copper alloys and alloys for electrical applications. The advantages of ESR over VAR were apparent when induction melting (IM) electrodes of lower-performance, heat-resistant alloys, such as Inconel 625 and Inconel 706, were remelted; these are used for nonrotating components of aircraft engines. In the next few years, test melting of heat-resistant alloys that are used for rotating engine components was performed in order to study the interaction between the slag and alloys containing active elements as well as segregation in larger ingots with a diameter of 430 mm or more. The high potential of ESR steel for production of turbojet rotating components was made clear by the simultaneous scientific and engineering developments in a number of different areas.

## Fracture Mechanics

Fracture mechanics, which used to be a laboratory discipline, has now became a useful tool for estimating the lifetime to components that determine the safety and reliability of airplanes. Methods for identifying small nonmetallic inclusions

---

This paper is printed in abridged form.
Authors' affiliations: J. Domingue, Special Metals Corporation, New Hartford, New York, USA; K.O, Yu, PKK Airfoils, Inc. Cleveland, Ohio, USA.

that are not detectable by ultrasonic inspection were developed in parallel with fracture mechanics. Thus, it became possible to replace general-purpose, heat-resistant alloys, which were no longer satisfactory, with more highly developed heat-resistant alloys.

## Rapid Increases in the Cost of Power and Raw Materials and Other Production Expenses

In order to reduce fuel costs, jet engines with a low specific power consumption were required, and it was therefore necessary to increase the heat resistance of the materials used in engines. In order to reduce raw material and production costs, engines with a longer lifetime and reduced operating and maintenance costs were desirable. Furthermore, steel production technologies had to be efficient. In this situation, it was not feasible to use more highly developed, heat-resistant alloys with high concentrations of scarce elements nor to employ complex technologies, such as powder metallurgy (PM), in spite of the latter's high efficiency and increased productivity, which are due to the ability to make parts with a shape close to the final product.

## Achievements in Microelectronics

The use of microelectronics promises to eliminate restrictions that limited the capability of furnace control systems. Because of these achievements, new comparative analyses of ESR and VAR are now justified, but on a commercial scale, ESR might improve steel quality in a shorter time than other advanced technologies, such as electron beam melting (EBM), plasma metallurgy, and PM. In addition, the use of microprocessors for data processing accelerates the assessment of the quality of ESR heat-resistant alloys.

The development of ESR occurred in several stages. At first, ESR was regarded as a way to produce high-quality electrodes for VAR that had a low concentration of inclusions [5]. In the next stage, ESR ingots were produced. For this purpose, it was necessary to take a new approach to the problem of heat flow during melting of a consumable electrode, which was always associated with VAR in the United States, and apply it to the conditions of the ESR process. The tendencies for inhomogeneity and segregation were also studied; these are problems that were not typically seen in traditional ESR alloys.

The heat-resistant alloy Inconel 718 is widely used in the United States and Western Europe for highly stressed rotating components, such as disks [2]. This alloy was studied to determine the applicability of ESR in aerospace engineering. We should mention that Inconel 718 is used in the for manufacture of a great number of components. However, the results of studies on Inconel alloys with lower heat-resistant characteristics could not be used to develop remelting conditions for Inconel 718. ESR techniques are now being developed for producing Waspalloy, which is another alloy used in disk manufacturing.

**Table 40.1.** Chemical composition of heat-resistant alloys (%)

| Alloy | C | Al | Ti | Nb | Cr | Co | Fe | Mo | Ni |
|---|---|---|---|---|---|---|---|---|---|
| | | | *Aircraft engine disks* | | | | | | |
| Inconel 718 | 0.04 | 0.5 | 1.0 | 5.3 | 18.1 | — | 17.5 | 3.0 | 54.6 |
| Waspalloy | 0.05 | 1.3 | 3.0 | — | 19.5 | 13.5 | — | 4.3 | 58.4 |
| | | | *Stationary aircraft engine components* | | | | | | |
| Inconel | | | | | | | | | |
| 625 | 0.03 | 0.2 | 0.2 | 3.6 | 22.0 | — | 2.0 | 9.0 | 63.0 |
| 706 | 0.04 | 0,3 | 1.8 | 2.9 | 16.0 | — | 37.3 | — | 41.7 |
| Hastalloy 750 | 0.04 | 0.8 | 2.5 | 1.0 | 15.5 | — | 7.0 | — | 73.2 |

A comparison of typical aloys used for disk manufacturing and alloys for re-ciprocating aircraft engine components is given in Table 40.1.

## Equipment

A detailed description of the ESR installation is given in [4]. The basic character-istics of the furnace are as follows: a stationary freezer mold with a round or a rectangular cross section, a maximum melting current up to 35 KA (DC or AC), and a maximum ingot weight of 18 tons. Recently, 760-mm diameter ingot weighing 10 tons was produced. A liquid start is performed by pouring molten slag from the top of the mold into an annular gap through a removable rectangu-lar slag tube. The installation is supplied with equipment to control the protec-tive atmosphere and with copper, water-cooled, coaxial current leads (water velocity, 4.6 m/s) that are insulated from the freezer mold to reduce magnetic interference.

Soon after the furnace was built, the joint between the freezer mold and the bottom plate was modified so as to provide an electrically conductive mold. This conductive freezer mold was used for investigations of heat flow by K.O. Yu and coworkers.

## Discussion

### Heat Balance and Characteristics of Heat Transfer

The results of studies of the heat balance in a commercial ESR furnace are discussed in the literature [6].[1] It was determined that about 87% of the heat that

---

[1] Analalogous results were obtained in earlier studies of soviet scientists, B.E. Paton and B.I. Medovar, "Thermal processes during electroslag remelting," Naukova dumka, Kiev (1979) [editor's note].

is generated is absorbed by the water-cooling system through the freezer mold walls, and 5% through the bottom plate. The remaining 8% of the heat is lost by radiation from the surface of the slag layer. Heat losses through the freezer mold water-cooling system, bottom plate, and slag layer surface are 44%, 23%, and 4%, respectively. The remaining 29% of the input energy are retained in the molten steel as heat of fusion and specific heat.

K.O. Yu and coworkers also developed a nonstationary heat-transfer model in which a sinusoidal variation of the latent heat of fusion was chosen for modeling ingot thermal conditions. The shape of the molten steel pool was determined by introducing tungsten powder into the melt when the ingot length to width ratio reached 1, 2, 3, and 4. There is good agreement between the calculated shapes of the solidification profiles and the length of the cylindrical part of the molten steel layer and the experimental data (Fig. 40.1). It is shown in Fig. 40.1 that a molten

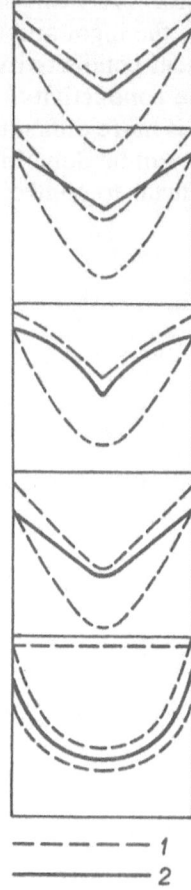

**Fig. 40.1.** Comparison of (1) calculated and (2) measured solidification fronts for ESR ingots from Inconel 718 with a diameter of 432 mm and a length of 2,007 mm.

metal depth between 203 and 299 mm, corresponding to a quasi-stationary process, was observed when the ingot length exceeded two diameters.

The structure of ESR and VAR ingots depends to a large extent on the thermal history of the casting. As a result of the high cooling rates, a thin molten metal layer is formed, with a narrow two-phase region, that minimizes segregation and improves the structural homogeneity of the ingot. In order to better understand the reasons for differences in the ingot structure, the thermal characteristics of the copper freezer mold were thoroughly studied for both ESR and VAR processes [7].

Heat transfer through the freezer mold in ESR and VAR is a complex process that depends on the following factors: the thickness of the slag skin (ESR); the helium pressure in the gap between the ingot and the mold, and the width of the gap (VAR); the cooling water flow rate; and the condition of the freezer mold surface. It has been shown [7] that the heat conductivity over most of the surface of VAR ingots is higher than that of ESR ingots, except at the top of the ingot (Fig. 40.2). The heat conductivity for VAR can be increased by increasing the helium pressure in the gap between the ingot and the mold. For ESR, the most important factor influencing the heat conductivity is the thickness of the slag skin, increases of which reduce the conductivity. Since the slag skin thickness increases at lower melting rates, it is more difficult to change its thickness independently of the melting power, as can be done in VAR by altering the helium pressure. Therefore, it is more difficult to control the cooling rate of ingots in ESR than in VAR.

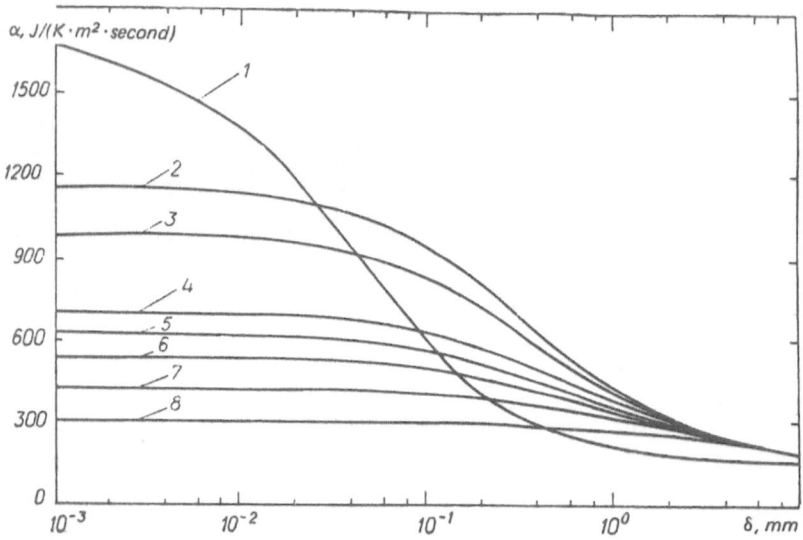

**Fig. 40.2.** Dependence of the heat transfer coefficient $\alpha$ on the width of the gap between the outer surface of ESR and VAR ingots and the mold side wall $\delta$ for different gas pressures: (1) ESR, air, 0.1 MPa; (2) to (8) VAR, helium pressures 3,990, 2,660, 1,330, 1,064, 798, 532, and 266 Pa.

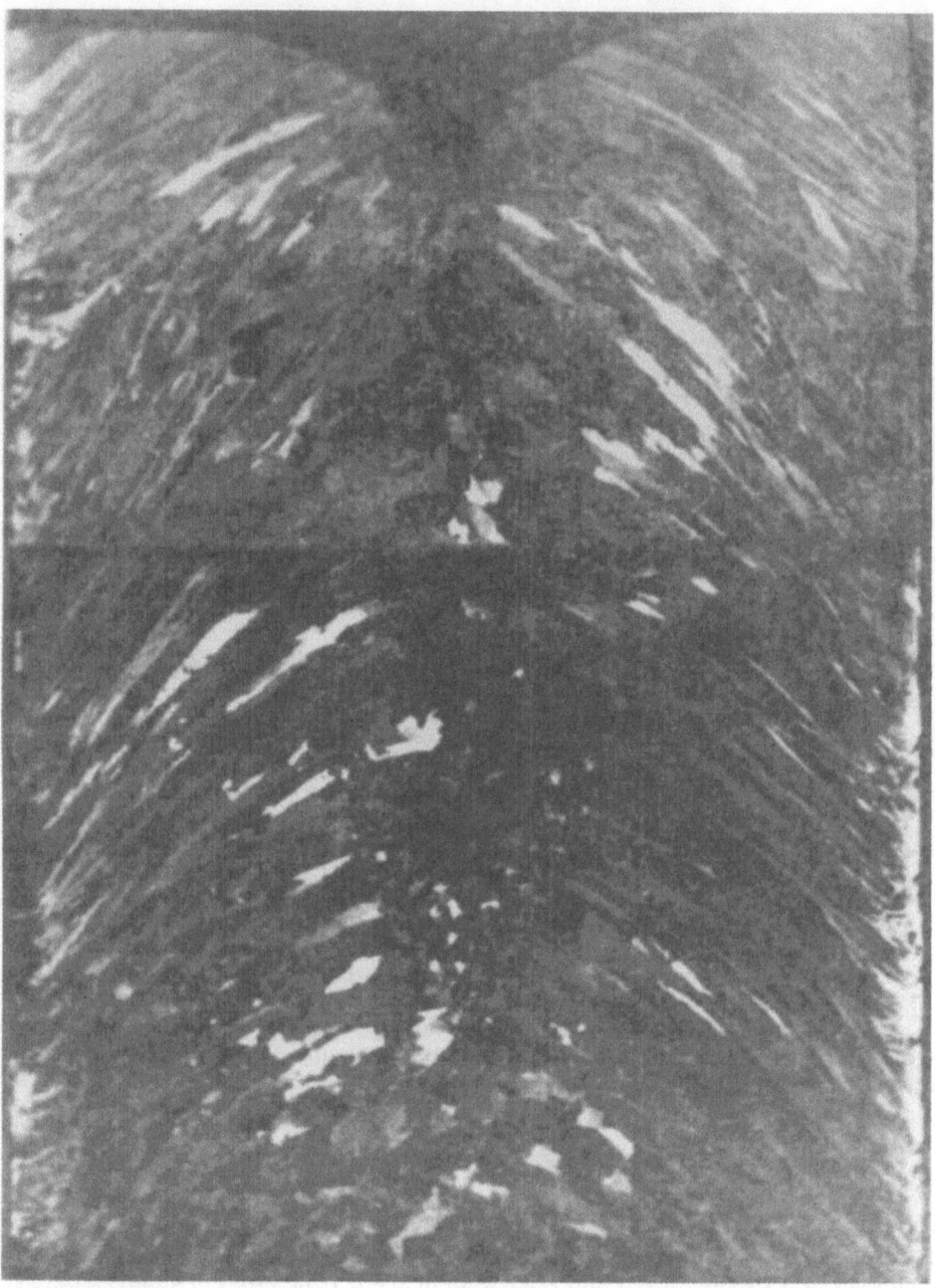

**Fig. 40.3.** Macrotemplet of the top end of an Inconel 718 ESR ingot with a diameter of 432 mm; melting rate was 591 kg/h.

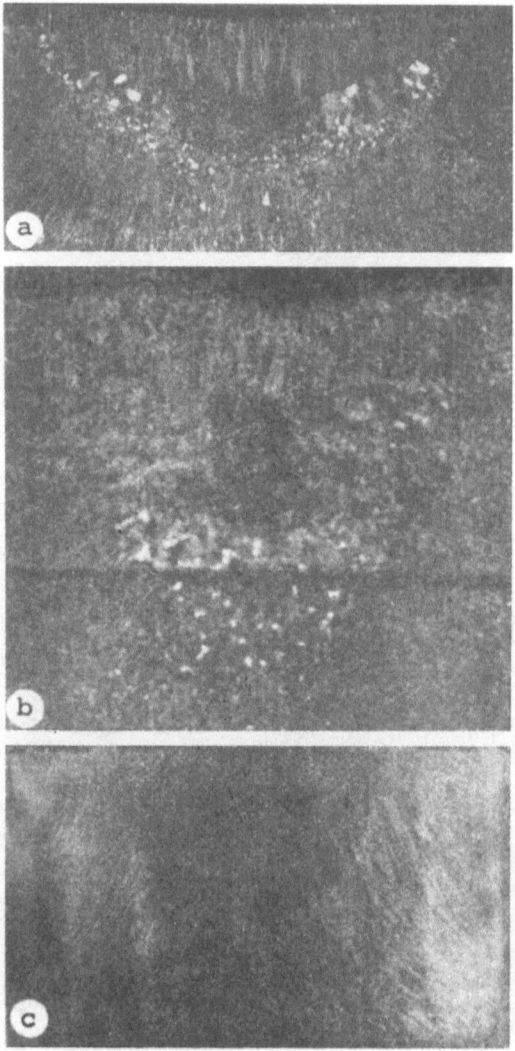

**Fig. 40.4.** Macrotemplets of the top ends of three Inconel 718 VAR ingots with a diameter of 508 mm: (a) melting rate 182 kg/h, no termination; (b) melting rate 322 kg/h, no termination; and (c) melting rate 355 kg/h, with termination.

## Ingot Structure

The influence of heat transfer in ESR and VAR on the metal pool shape and on the grain size have been studied [8]. It was determined that the metal pool is V-shaped in ESR (Fig. 40.3) and U-shaped in VAR (Fig. 40.4). The direction of grain growth in ESR in the vicinity of the ingot surface is almost parallel to the vertical axis and is angled 45° from the vertical in the center of the ingot.

The direction of grain growth in VAR ingots is 90° to the vertical in the vicinity of the surface and is almost vertical in the center (Fig. 40.5).

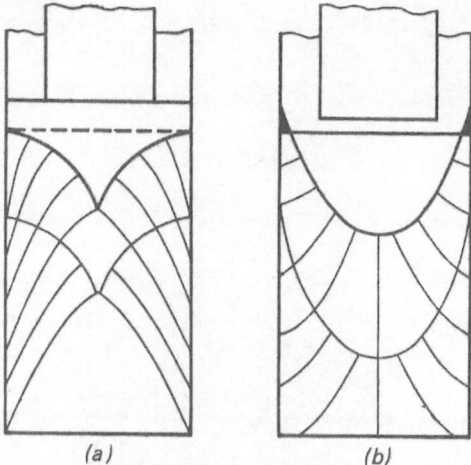

**Fig. 40.5.** Influence of heat flow on the shape of the molten metal pool and on the grain orientation for (a) ESR and (b) VAR.

Studies of the grain size and the distance between dendrite branches in the radial direction showed that the peripheral regions of ESR ingots cool significantly faster than the central part. The difference in cooling rates and the difference in the distance between the dendrite branches of the central and peripheral regions of VAR ingots are relatively small. An estimate [6] of the width and shape of the two-phase zone in ESR and VAR ingots showed that the V-shaped solidification front in ESR enlarged the two-phase zone in the center of the ingot; that is, the two-phase zone is nonuniform across the ingot section. VAR ingots, which have a U-shaped solidification front, have a narrower, relatively even distribution of the two-phase zone along their cross sections (Fig. 40.6).

## Results

The first studies of ESR of Inconel 718 in freezer molds with diameters of 432 and 508 mm (before 1982) yielded forgings with an unsatisfactory macrostructure. The main problems were ghost segregation defects and dispersion harden-

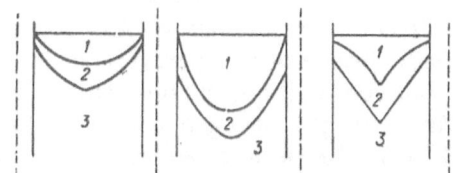

**Fig. 40.6.** Calculated dimensions of the two-phase zone in ESR and VAR Inconel 718 ingots: (1) molten steel; (2) two-phase zone; and (3) solid alloy.

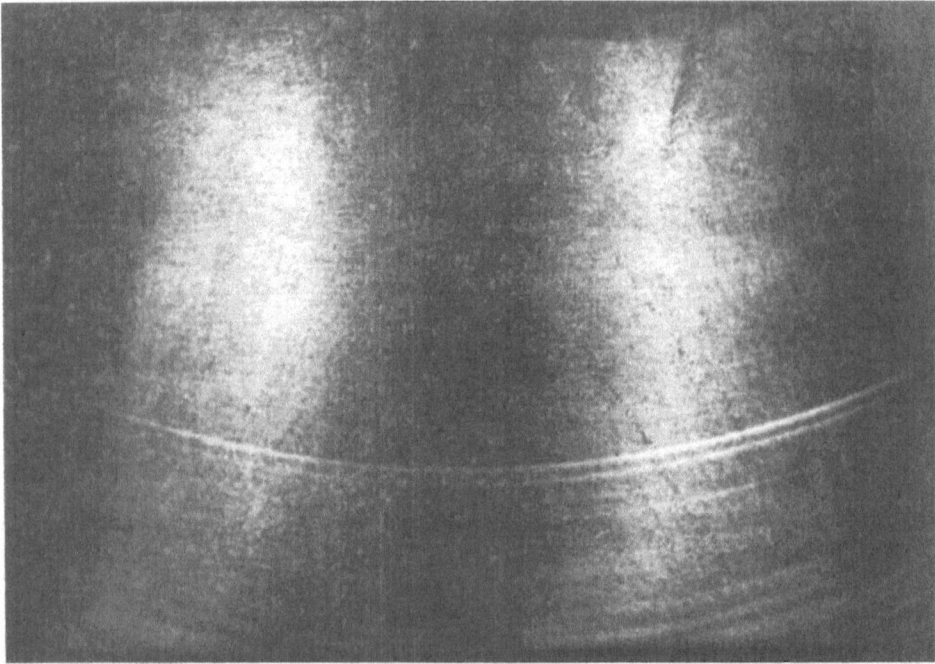

**Fig. 40.7.** Longitudinal templet of an ingot with a diameter of 508 mm.

ing defects. Ghost segregation was revealed by an etch sensitive to niobium (Fig. 40.7) [1]. At that time, the remelting technique was based on melting conditions that were developed for specialized steels, and provided good results for heat-resistant alloys with a low concentration of hardening elements. In the next year, ESR of Inconel 718 with a very low nonmetallic inclusion content predetermined further investigations in this field.

The identification of nonmetallic inclusions was verified by electron-beam melting of test "pellets" weighing from 0.5 to 1 kg.[2] The surface of an inclusion-free pellet is shown in Fig. 40.8c [9, 10]. Test pellets produced by IM and IM + VAR contained agglomerated particles of titanium nitride on shrinkage pipes. These studies proved that ESR is one of the most advanced methods for the production of high-purity, heat-resistant alloys. Furthermore, the ESR ingot varied in chemical composition only in the bottom and top ends.

The studies of steel purity were an important turning point in the development of remelting techniques for heat-resistant alloys. As a result, conventional techniques for ESR of specialized steels were replaced by more advanced methods in which the melting rate, slag composition, and protective atmosphere were varied to minimize the reactivity of oxygen, nitrogen, and silicon (all these elements are undesirable in Inconel 718). The desired parameters for obtaining pure alloy had to be coordinated with the results of heat balance studies. The dominant factor determining the heat transfer coefficient is the slag skin thickness [3, 7]. Insufficiently high melting rates result in the formation of a thick slag skin,

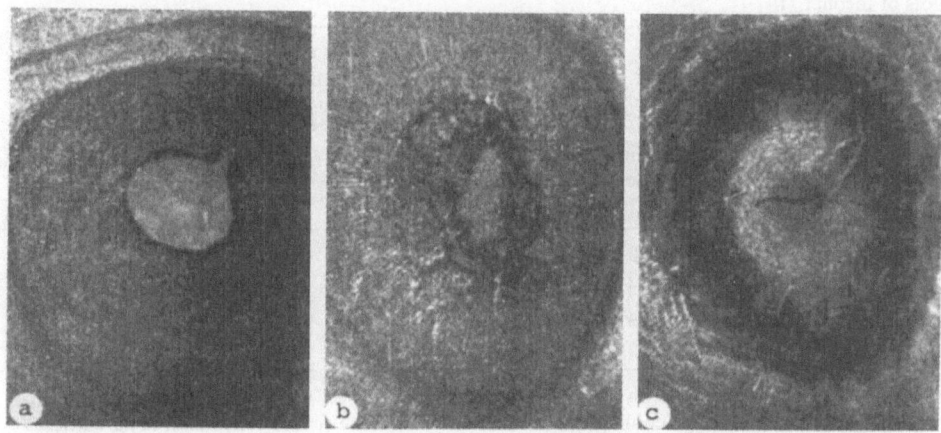

**Fig. 40.8.** Test "pellets" of Inconel 718 produced by (a) IM, (b) IM + VAR, and (c) IM + ESR after electron-beam melting.

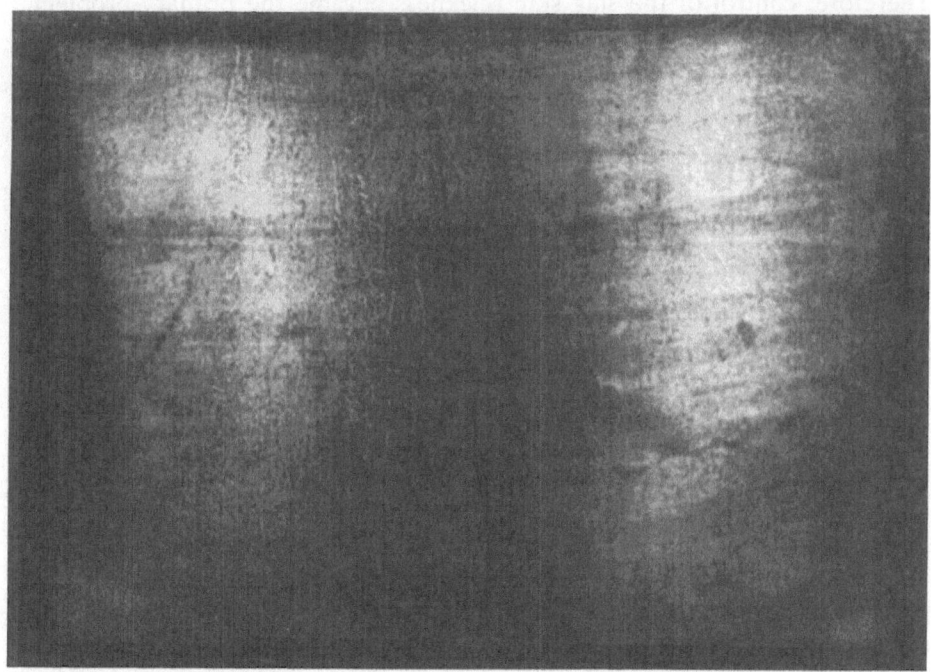

**Fig. 40.9.** Longitudinal templet of an ESR Inconel 718 ingot.

**Table 40.2.** The width of the two-phase zone in electroslag-remelted and vacuum arc-remelted ingots of Inconel 718

| Production technique | Melting power (KW) | Melting rate (kg/h) | Ingot diameter (mm) | Distance between dendrite branches ($\mu$m) | | Width of two-phase zone (mm) | | Shape of the two phase zone (see Fig. 40.6) |
|---|---|---|---|---|---|---|---|---|
| | | | | C | E | C | E | |
| VAR | 125 | 180 | 508 | 131 | 101 | 123 | 57 | a |
| | 200 | 322 | 508 | 114 | 104 | 146 | 115 | b |
| ESR | 240 | 273 | 432 | 113 | 74 | 167 (152) | 48 (40) | c |

Note: Calculations were based on the equation $d = 33.85\,\varepsilon^{-0.338}$; $d$, distance between dendrite branches ($\mu$m); and $\varepsilon$, average cooling rate of the two-phase zone (cal/sc). The data in parentheses were taken from preliminary thermal calculations.

which prolongs the solidification period and causes dendrite segregation and whiskers (Fig. 40.9) [2]. Unfortunately, a thin slag skin is obtained at higher melting rates, which are also accompanied by ghost segregation.

Therefore, control of the slag skin thickness became the primary problem. The composition of a slag consisting of 70% $CaF_2$, 15% $CaO$, and 15% $Al_2O_3$, which is used to produce low-performance heat-resistant alloys, was corrected. An inert gas protective atmosphere provided the needed purity and chemical composition along with the required ingot structure. The results of these studies show that the optimal melting rates for ESR are comparable to those of VAR. A semifinished product forged from an ingot that was produced in a 432-mm diameter freezer mold had a uniform structure [1–3]. A similar forging from Waspalloy was also free from segregation.

## Future Outlook for ESR

The greatest attention is now focused on obtaining reliable data for ESR of Inconel 718 in 432-mm diameter freezer molds in order to compare them to data gathered over a period of nearly 25 years. In order to estimate the quality of rotating components for aircraft engines fabricated from ESR steel, a complex mathematical model along with control systems as required.

At present, ESR techniques are being developed for the production of Inconel 718 ingots with a diameter of 508 mm, which are being produced primarily by VAR.

[2] A detailed description of the technique is given in "The present state of electric arc melting," *Electroslag Remelting 9*, Naukova dumka, Kiev (1987).

# 41 Twenty-Five Years' Experience in Electroslag Remelting Development

## J.B. Rambo

The first electroslag (ES) furnace in Western Europe was put in operation with the help of the E.O Paton Electric Welding Institute (Kiev, USSR) In July 1965 at a plant in Firmini (France).[1] In 1973, at the IV Aircraft Conference in Burge, the first report comparing the characteristics of electroslag-remelted (ESR) and vacuum arc-remelted (VAR) steels for landing gear components of the "Concord" and "Aerobus" aircraft was presented. At the Aircraft Conference in June 1981, a report was made on investigations of the quality of critical aircraft components fabricated from steel produced by improved ESR techniques and equipment.

This chapter, in abridged form, summarizes ESR development at KZF, based on an analysis of various types of production.

Historically, VAR was the first method used to obtain high-quality products; it was used in Europe and in the United States. Originally, ESR was used as a rather simple method for obtaining low-sulfur steel and as a more economical method than VAR for increasing steel purity.

Initially, efforts were made to use a single ESR process to produce steel of a quality that, at the time, could be a guaranteed only by VAR, especially for such characteristics as micropurity, fatigue strength, and mechanical properties. Through a general research program, significant success was achieved in meeting the growing needs for high-quality steels for the aircraft and power industries. Priority was placed on two quality levels of products: special-quality products used in aircraft construction and extra-high-quality products used in the construction of highly reliable aircraft components.

This chapter presents some of the test results obtained for various types of components.

## Equipment

At present, it is possible to obtain four types of ingots without replacement of electrodes: square cross section with rounded corners measuring $450 \times 450$ mm and weighing a maximum of 5 tons; round with a diameter of 700 mm and

---

[1] In 1963 the Firminy Plant belonged to the "Compaignie des Ateliers et Forges des la Loire" (CAFL) [editor's note].

---

Author's affiliation: Forgings and Castings Production Company, KZF, Usinor Group, France.

weighing a maximum of 10 tons; round with a diameter of 850 mm and weighing a maximum of 15 tons; and round with diameter of 1.1 m and weighing a maximum weight of 26 tons.

The weight of circular ingots with a diameter of 1.1 m can be increased to 33 tons by using electrode replacement during remelting. The movable bottom plate that is used for 33-ton ingots can also be used with equipment for melting hollow ingots, if necessary. The furnace is connected to a computer that provides automatic control of remelting. The computer also keeps track of the electrode weight and the electrical parameters through a real-time interface. This ensures that the melting rate will be held to the programmed value.

## Control of the Atmosphere Above the Slag Layer

Different installations allow for a dry and controlled atmosphere or an atmosphere with a limited and controlled volume. Remelting can be initiated either with a solid flux charge or with a liquid charge that is melted in an auxiliary furnace.

## Ingot Characteristics

### Requirements for the Ingot Surface

As is well known, the slag skin that forms between the ingot and the freezer mold during remelting results in a smooth ingot surface. With a stationary freezer mold, the results are evident from the very beginning, whereas for 1.1-m diameter ingots that are withdrawn from the mold, appropriate selection of the slag (viscosity, hardness, etc.) and the withdrawal method is very important.

### Homogeneity of Chemical Composition

For applications in which a uniform distribution of alloying elements is essential, and especially of oxidation-susceptible elements, efforts were made to reduce the differences in chemical composition between the top and bottom of the ingot by selecting special slags and remelting conditions.

### Absence of Internal Defects

Long experience with ESR has confirmed the two main advantages of the process: the ability to increase steel purity and the ability to improve the crystal structure of the ingot. After ESR, the large inclusions that were present in the consumable electrode are no longer detectable. The inclusions are removed when the steel passes through the slag layer. Desulfurization provided by proper deoxidation of the slag results in the reduction or even the complete elimination of sulfides. Proper selection of the slag can change the inclusion size and type and thus reduce their detrimental effects. The remaining inclusions are relatively small and are classed as type D (spheroidal oxides). The combination of intensive cooling from the freezer mold and the low solidification rate of the ingot is another advantage of using ESR; it produces homogeneous ingots that are free from axial segregation defects.

## Mechanical Properties

The crystal structure of the steel as well as its purity improve such mechanical properties as impact resistance, ductility, fatigue strength, and crack resistance. The homogeneity of ESR steel is largely responsible for the isotropy of its mechanical properties in the transverse and the longitudinal directions.

## Special Steels Used in Aircraft Engineering

A steel quality sufficiently high to guarantee high service reliability can be obtained, provided that inclusions are assessed according to the modified standard ASTM E 45. According to this specification, the number of inclusions should be under 1.5 for each field of view, and all the inclusions should be type D (spheroidal oxides).

The total number of inclusions is calculated by adding up the number of type A, B, C, and D inclusions detected in 40 fields of view; there should be less than 35 of each type. The dependence of the total number of inclusions on their size and the steel production technique is given in Fig. 41.1.

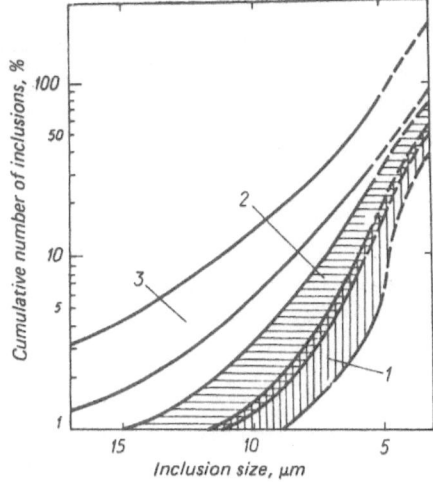

**Fig. 41.1.** Cumulative distribution of nonmetallic inclusions by size: (1) ESR steel, extra high-quality grade; (2) ESR steel, regular aircraft grade; and (3) initial alloy (according to ASTM E 45, 40 fields of view).

226                                                                J.B. Rambo

Inclusions are grouped according to the reduction in their size. Compared
with the steel conventionally used in aircraft engineering, special-quality ESR
steels have a greatly reduced number of inclusions with a diameter greater than
20 $\mu$m.

ESR improves the homogeneity of steel, providing almost isotropic mechanical properties, except for slight differences in the reduction of area of transverse
samples and in the impact resistance of transverse and longitudinal samples.

A comparative analysis of the mechanical properties of six rotors fabricated
from Ni–Cr–Mo–V steel, of which three were made from ESR steel and three
from regular steel of the same heat, showed a remarkable improvement of the
properties of the steel, such as elongation, impact resistance, and ductile–to–
brittle transition temperature. The improvement in the ductility observed in
nonmagnetic turbine rings fabricated of 18Mn5Cr steel is of particular interest.

## Extra High-Quality ESR Steel

Steels used for manufacturing critical aircraft components that function under
conditions of extreme stress should be of exceptionally high quality. This type of
specification was used for the first time in France for landing gear of the "Concord" airplane, and later for "Aerobus" A.300 landing gear fabricated of VAR

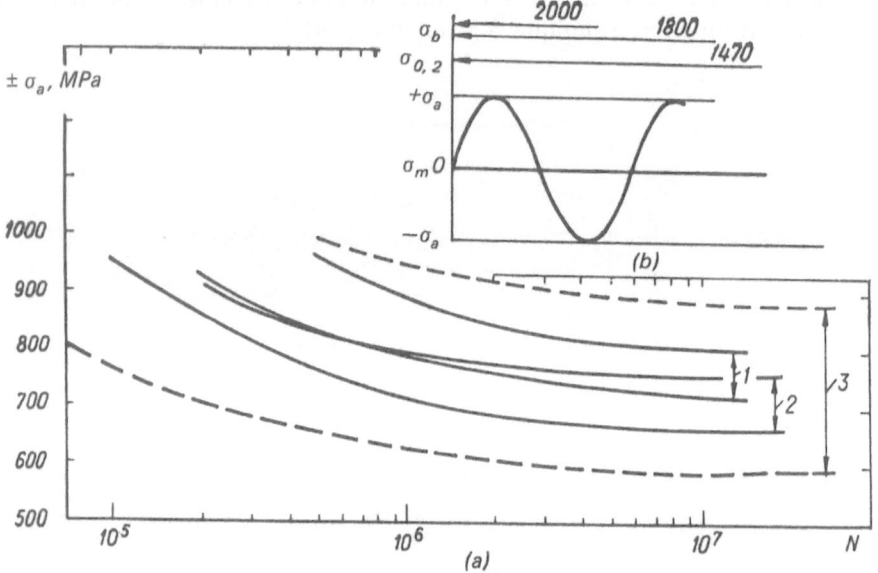

**Fig. 41.2.** (a) Cyclic fatigue resistance; and (b) applied stress and mechanical properties of samples:
(1) and (2) longitudinal and transverse specimens; and (3) transverse VAR specimens ($N$, number of
cycles).

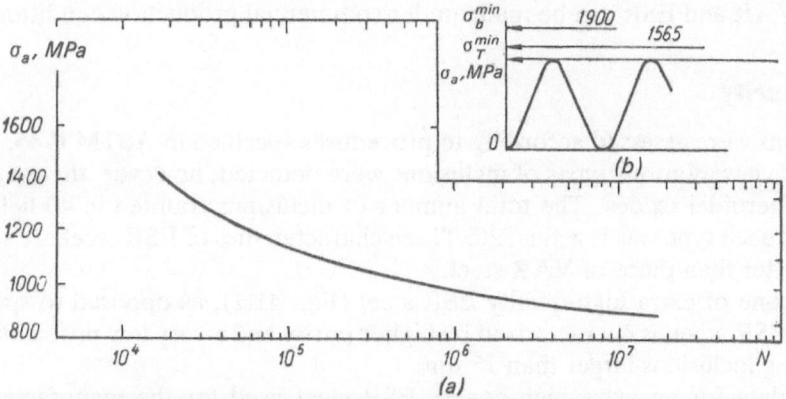

**Fig. 41.3.** (a) Cyclic fatigue resistance; and (b) applied stress and mechanical properties of 434 OM ESR steel (transverse samples were analyzed; $N$, number of cycles).

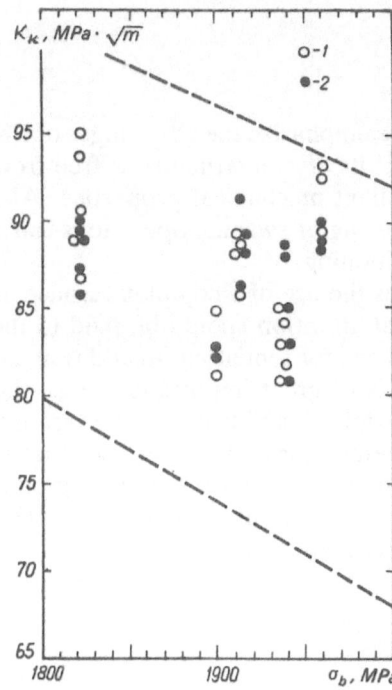

**Fig. 41.4.** Dependence of fracture resistance $\kappa_{IC}$ on tensile strength $\sigma_v$: (1) and (2) longitudinal and transverse samples; (dashed line) region of possible values

35NCD166 steel. Since there were many heats of this steel, a statistical comparison of VAR and ESR can be made under commercial production conditions.

## Micropurity

Inclusions were assessed according to procedures specified in ASTM E 45, in 40 fields of view. Various types of inclusions were detected; however, the majority were spheroidal oxides. The total number of inclusions counted in 40 fields of view for each type was less than 20. These characteristics of ESR steel are somewhat better than those of VAR steel.

The zone of extra high-quality ESR steel (Fig. 41.1), as opposed to special-quality ESR steel, is characterized by higher purity and a very low probability of observing inclusions larger than 15 $\mu$m.

Test data for an extra high-quality ESR steel used for the manufacture of highly reliable aircraft components are given in Fig. 41.2. These test results are the same as the results for VAR 35NCD166 steel. The test results for extra high-quality ESR 4340M steel are shown in Fig. 41.3. The characteristics for extra high-quality steel 35NCD166 produced by ESR and VAR are given in Fig. 41.4.

## Conclusions

The results presented here emphasize the advantages of ESR in comparison with VAR. ESR steel is clean, its crystal structure is free from defects, it is homogeneous, and it has excellent mechanical properties. Also, due to the surface finish of ESR ingots, reducing or swaging operations can be performed without preliminary surface conditioning.

However, ESR requires the use of a complex furnace installation with a controlled atmosphere. Great attention should be paid to the methods of refining and casting of electrodes used for remelting. In addition, an appropriate selecton of remelting parameters is of great importance (e.g., the slag should be very pure and have a high deoxidizing ability), as is the termination of remelting.

The optimization of metallurgical characteristics that makes ESR steel economical is of particular interest in connection with producion steel that can be used under various service conditions. This would make it possible to replace expensive VAR by cheaper ESR.

# 42 Parameters Influencing the Quality of Electroslag Ingots and Process Efficiency

R. Thielmann and J. Kreinenberg

The quality of electroslag-remelted (ESR) products and the economic efficiency of the process are determined by a great number of factors [1–4]. An analysis of these factors as presented in this chapter,[1] has led to the determination of two main directions for the further development of ESR. These are improving the quality of high-alloy and precision steel ingots, on the one hand, and increasing the productivity of remelting large diameter ingots of conventional types of steel, on the other.

An experimental ESR installation was used to study the interdependence of the following parameters (independent of further ingot processing): the solidification rate, the depth of electrode penetration, the type and volume of the slag, the mold filling factor, termination of remelting, specific power consumption, ingot surface quality, the shape of the molten metal pool, shrinkage defects, and the chemical composition of the steel [5–9].

## Experiments

Electrodes with a diameter of up to 220 mm were remelted into ingots with a diameter of 300 mm. Remelting was done in air, vacuum, and an inert gas atmosphere at a pressure of $3 \times 10^5$ pa, using direct and alternating current (frequency from 1 to 50 Hz), and a maximum voltage and current of 100 V and 10 kA, respectively.

The process was controlled by a digital computer system. The control system was developed to regulate the melting rate and depth of electrode penetration, which aided the experimental work. A high-precision weighing system was developed to select the appropriate melting rate according to the electrode weight (Fig. 42.1).

Five different steel types (Table 42.1) and four slag compositions (Table 42.2) were analyzed. For example, the parameters for one of the melting runs were as follows: steel, 21; electrode weight, 320 kg; electrode diameter, 170 mm; ingot diameter, 300 mm; pressure above the slag layer, $10^5$ pa; power supply

---

[1] Abridged from a paper presented at the VIII International Conference on Vacuum Metallurgy, Linz, Austria, September 30 to October 4, 1985.

---

Authors' affiliations: R. Thielmann, Thiessen Edelstalwerke AG, Krefeld, West Germany; J. Kreinenberg, Betriebsforschungsinstitut, Dusseldorf, West Germany.

**Table 42.1.** Chemical composition of the remelted electrode (%)

| Steel type DIN | AISI | C | Si | Mn | P | S | Cr | Mo | Ni | V | W | Al | B | N | Ti | O |
|---|---|---|---|---|---|---|---|---|---|---|---|---|---|---|---|---|
| 21NiCrMoS$_2$F | 8,620 | 0.21 | 0.24 | 0.75 | 0.017 | 0.027 | 0.62 | 0.23 | 0.55 | 0.01 | 0.01 | 0.026 | 0.004 | 0.009 | 0.01 | 0.005 |
| 17NiCrMo$_6$ | — | 0.17 | 0.25 | 0.52 | 0.014 | 0.025 | 1.70 | 0.28 | 1.40 | 0.01 | 0.01 | 0.033 | 0.004 | 0.01 | 0.01 | 0.009 |
| 100Cr6 | 52,100 | 0.91 | 0.40 | 0.99 | 0.006 | 0.006 | 1.45 | 0.01 | 0.07 | 0.01 | 0.01 | 0.013 | — | 0.005 | 0.01 | 0.020 |
| X2CrNiMoTi216 | 329 | 0.04 | 0.38 | 0.49 | 0.026 | 0.003 | 20.92 | 2.01 | 5.88 | 0.09 | 0.07 | 0.029 | 0.0046 | 0.039 | 0.3 | 0.005 |
| S6-5-2 | | 0.91 | 0.37 | 0.29 | 0.024 | 0.015 | 4.19 | 4.69 | 0.21 | 1.79 | 6.05 | 0.005 | 0.0013 | 0.021 | 0.01 | 0.005 |

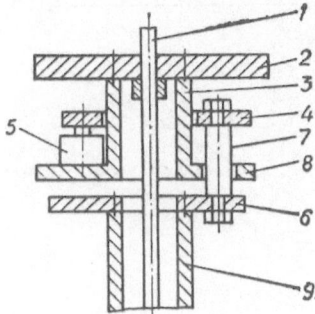

**Fig. 42.1.** Diagram of the weight measuring system of the experimental furnace: (1) connecting rod; (2) cross-member; (3) nut; (4) upper flange; (5) load cell; (6) base plate; (7) studs; (8) lower flange; and (9) electrode suspension.

**Table 42.2.** Chemical composition of slags used for electroslag remelting (%)

| Slag type | $CaF_2$ | CaO | $Al_2O_3$ | $SiO_2$ | MgO | FeO |
|---|---|---|---|---|---|---|
| 60/20/20 | 54.84 | 18.18 | 20.18 | 1.70 | 0.24 | 1.05 |
| 0/50/50 | 0.15 | 45.77 | 50.23 | 0.51 | 0.46 | 2.23 |
| 0/55/32/7/6 | 0.13 | 50.75 | 32.51 | 7.85 | 5.85 | 2.47 |
| 0/40/20/40 | 0.12 | 36.22 | 20.04 | 39.86 | 0.38 | 1.92 |

frequency, 2.5 Hz; solidification rate, 160 kg/h; water consumption, 12.5 m³/h; and slag weight, 15 kg (60% $CaF_2$–20% CaO–20% $Al_2O_3$).

This analysis reports only phenomenological dependences.

## Results

### Specific Power Consumption

Specific power consumption depends to a large extent on the ratio of the slag's heat-absorbing surface to its weight, that is, its quantity (Fig. 42.2). Reducing slag volume reduces power consumption to a greater degree than has been described in the literature [6, 15]. For the geometry and cooling conditions used in our experiments, the energy balance consisted of the follows components:

| Components | (kW-h/kg) |
|---|---|
| Heat input to melt the electrode | 0.40 |
| Thermal radiation losses from the slag surface necessary to overheat the steel droplets | 0.30 |
| Convection losses at normal atmospheric pressure | 0.10 |
| Heat losses from the slag to the freezer mold (slag layer depth, 100 mm) | 0.80 |
| Total power consumption | 1.60 |

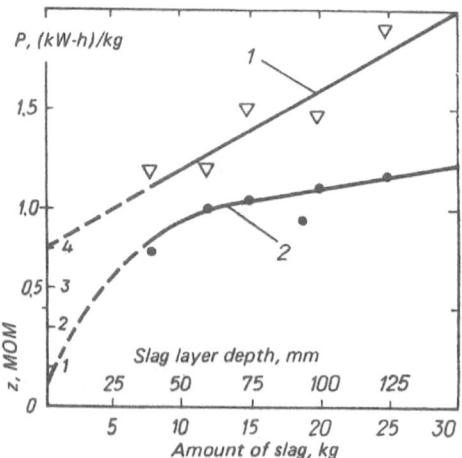

**Fig. 42.2.** Dependence of specific power consumption $P$ (1) and slag layer impedance $z$ (2) on slag layer depth.

The heat loss from the slag to the freezer mold amounts to a heat flow of 1,350 kW/m² (1.16 × 10⁶ kcal/m²-h), which is in good agreement with data for freezer molds of continuous ingot casting machines [17].

It was determined that the minimum power consumption corresponds to an electrode penetration depth of 15 mm. The voltage between the electrode and the ingot depends nonlinearly on the distance between the electrode and ingot (Fig. 42.3) [18]. With increasing electrode penetration depth, both the voltage and the resistance decrease greatly when the electrode end approaches either the metal pool or the slag layer surface. This dependence on voltage drop was calculated from changes in the current density that result from the changes in the

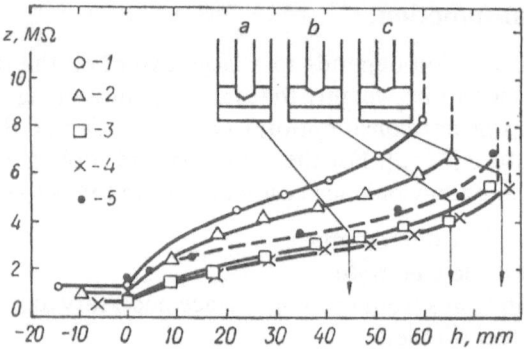

**Fig. 42.3.** Dependence of impedance ($z$) on electrode penetration depth ($h$): (1) alternating current, 2.6 Hz, 0/50/50 slag; (2) alternating current, 2.5 Hz, 0/55/32/7/6 slag; (3) reverse polarity direct current (electrode positive), 60/20/20 slag; (4) normal polarity direct current (electrode negative), 60/20/20 slage; (5) alternating current, 2.5 Hz, 60/20/20 slag; (dotted line) calculated dependence.

geometry. Conditions for minimum power consumption correspond to the maximum linear dependence between the voltage and the electrode penetration depth.

The specific power consumption depends on the slag composition [7, 9, 16, 19–21] and also on the slag layer resistivity and the potential drop at the slag–metal interface. Power consumption decreases with increasing slag resistivity, and at high resistivities, the power consumption reaches a constant value.

In high-resistivity slags, the convective flow is reduced, and sometimes it even changes direction. The combined influence of gravity and electromagnetic forces produces two toroidal circulating flows. Under normal conditions, external electromagnetic forces dominate [22], whereas for high-resistivity slags (i.e., in the case of high voltage drop, low current, and a weaker electromagnetic field), centrifugal convective heat flows become visible on the slag surface. The reduced (or even completely suppressed) convection exhibited by these slags, together with a higher viscosity and a lower heat conductivity [20, 23, 24], leads to reduced energy losses due to lower convective heat transfer within the bulk phase and lower heat exchange at the interface. To take full advantage of high-resistivity slags and taking into account power costs and capital investment, low current furnaces are required. In addition, in some cases this would permit decreasing the electrode penetration depth and increasing the remelting rate because convection and segregation in the molten steel are reduced also.

The dependence of the shape factor of the molten metal pool on the melting rate and the electrode penetration depth are given in Fig. 42.4 for a low mold fill factor and a low degree of cooling.

The formation of shrinkage pipes in ESR steel can be and should be prevented, although it is not always possible to do so in commercial production in which the time for termination of remelting is limited. In the final analysis, an optimal metal pool shape and the elimination of shrinkage pipes are achieved when remelting is performed under conditions of high thermal efficiency. When direct current (especially reverse polarity) is used for remelting, without termination, shrinkage defects are more prevalent. Slag composition has a signif-

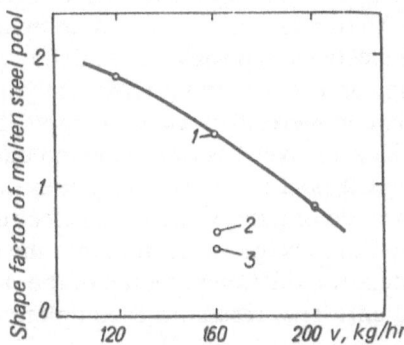

**Fig. 42.4.** The dependence of the shape factor of the molten steel pool on melting rate (v) for electrode penetration depth of (1) 0 mm; (2) 10 mm; and (3) 30 mm.

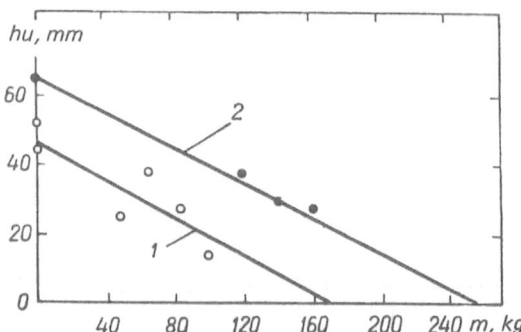

**Fig. 42.5.** The dependence of the length of the secondary shrinkage pipe (h) on the annealing time at termination of remelting for (1) 20 kg and (2) 15 kg of the slag; (m) weight of remaining electrode at termination of remelting.

icant effect on the formation of shrinkage defects. For identical melting rates, the depth of the shrinkage pipe was reduced by 30% when a 0/50/50 slag with no fluorides was used. When a $0/55/32/7SiO_2/6MgO$ slag was used, neither primary nor secondary shrinkage defects were observed, although the shape of the metal pool was not changed. This can probably be explained by the higher heat capacity of those slags [20, 23, 24].

The influence of annealing time after termination for a 60/20/20 slag on the length of the secondary shrinkage pipe is givn in Fig. 42.5. These results show that for small ingots it is not necessary to reduce the melting rate continuously when terminating remelting. In order to prevent the formation of shrinkage defects, it is sufficient to maintain the slag temperature close to the liquidus temperature of the steel.

## Oxide Content

Ingot purity, as measured by oxide content, is mainly determined by the type of current used for remelting [10, 11, 32]. Direct current remelting with the electrode negative (cathode) reduces ingot purity in comparison to the electrode alloy, whereas when the electrode is positive (the anode), purity does not change significantly. Alternating current remelting significantly increases purity. These results were confirmed in commercially produced ingots. It was determined that the atmosphere above the slag layer has practically no effect on ingot purity. In this respect, disregarding the polarity leads to an undesirable unidirectional flux of Fe(II) ions. The polarity determines whether the electrode or the molten steel surface is oxidized by the slag. Oxygen migration into the slag is markedly lower at the cathode, due to the low oxidation potential of the steel and the correspondingly low activity gradient. This results in a complete oxidation of the steel during remelting.

When alternating current is used for remelting (even for low frequencies), oxygen penetration is no greater than the thickness of the diffusion layer [14], which prevents oxygen transport.

The effect of slag composition on oxide content is insignificant. Even a slag containing 7% $SiO_2$ can give very good results, as reported in the literature [9, 12], provided that $SiO_2$ activity is sufficiently low [23].

## Chemical Composition

Besides reactions with sulfur and hydrogen, reactions with oxygen are of particular significance. The oxygen concentration in the steel is determined by reduction–oxidation reactions, in which slag composition is in equilibrium with oxygen activity [33, 34].

In our studies, there was no significant effect of slag composition on total oxygen concentration in the ingot (Fig. 42.6). Oxygen concentration is very low even when a slag containing $SiO_2$ is used. The concentration of aluminum is determined only by the $Al_2O_3$ content of the slag. Normal polarity direct current remelting (i.e., electrode negative) (Fig. 42.7) results in the highest oxygen concentration in the steel (200% higher than the oxygen concentration in the electrode alloy), whereas with 50-Hz alternating current, the oxygen concentration is reduced by 60% to 80%. This is in partial disagreement with data in the literature [4, 12], where the contradictory results are explained as being due to differences in ingot diameter and current density. In an inert gas atmosphere, oxygen concentration is somewhat lower than it is during open air remelting (Fig. 42.8).

The type of current (DC or AC) does not affect the concentration of other elements. As was expected, the reduction in silicon concentration for open air remelting is twice as great as for inert gas remelting. This process depends to a large extent on the content of aluminum, which is a strong reducing agent. In contrast to open air remelting, which decreases the silicon concentration,

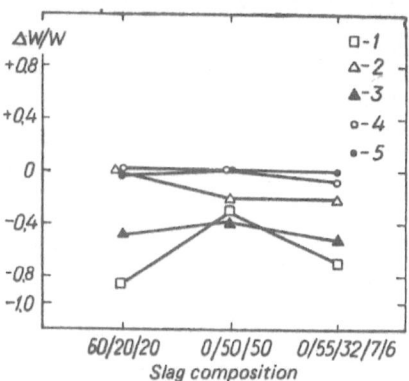

**Fig. 42.6.** Fractional change in the concentration of various elements ($\Delta W/W$). (1) Al; (2) O; (3) Si; (4) C; and (5) Mn, in $21NiCrMoS_2F$ alloy after remelting in air using various composition slags and electrodes containing 0.025% Al; 0.005% O; 0.24% Si; 0.21% C; 0.73% Mn.

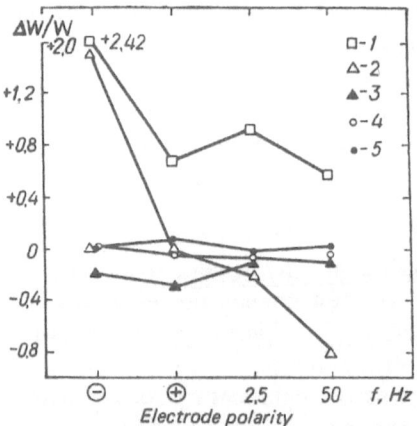

**Fig. 42.7.** Fractional change in element concentration ($\Delta W/W$). (1) Al; (2) O; (3) Si; (4) C; and (5) Mn, after remelting 21NiCrMoS$_2$F alloy electrodes containing 0.026% Al; 0.005% O; 0.24% Si; 0.21% C; 0.73% Mn, using various electrical parameters under a reduced pressure of argon; (f) AC frequency.

remelting in an inert gas atmosphere using an aluminum-bearing 60/20/20 slag increases the silicon concentration by 80%.

## Process Control

The melting rate was controlled through an HP-1000 computer, which maintained the current and voltage setpoints. The furnace control system has a

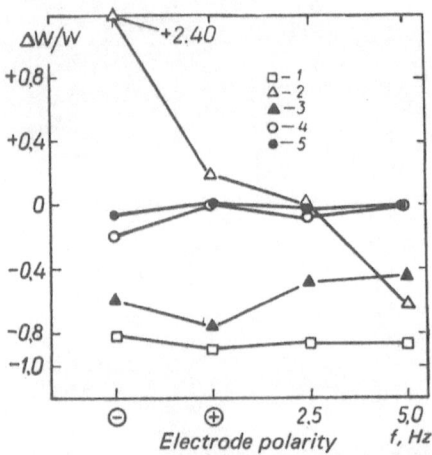

**Fig. 42.8.** Fractional change of element concentration ($\Delta W/W$) (1) Al; (2) O; (3) Si; (4) C; and (5) Mn, for electrodes of 21NiCrMoS$_2$F alloy containing 0.026% Al; 0.005% O; 0.24% Si; 0.21% C; 0.73% Mn, after remelting in air with various electrical parameters; (f) AC frequency.

hierarchial structure that decreases the reaction time for the control of the current, voltage, and electrode penetration depth.

The effect of certain remelting conditions on the efficiency and product quality was studied at the experimental ESR plant. It was determined that AC frequency, type of atmosphere, and slag composition are the most important process parameters. Their effect on specific power consumption, steel purity, and the shape of the metal pool was studied.

Alternating current remelting led, in most cases, to a reduction of specific power consumption (to 1.5 kW-h/kg) in comparison with direct current remelting (up to 2.3 kW-h/kg). For this purpose, it is preferable to use normal polarity (electrode negative) when performing remelting in an air atmosphere. This dependence of power consumption is weaker for remelting in an inert gas atmosphere.

In experiments in which the melting rate was kept constant, the shape of the metal pool and solidification conditions are of great importance from the point of view of reducing specific power consumption.

The advantage of high-resistance slags is emphasized. The use of AC affects the oxide content of the ingot, independent of the frequency and the type of atmosphere above the slag layer.

With reversed polarity direct current remelting (electrode positive), the ingot purity differs slightly from the electrode purity, whereas for normal polarity current, the purity is significantly degraded, especially in an inert gas atmosphere. Digital control of the process was by a computer. The control system maintains the melting rate and the electrode penetration depth at a given programmed value. For accurate control of the melting rate, a high-precision electrode weighing system was developed. This closed-loop control system can also be used under commercial production conditions.

# References

1   A. Mitchel and R.M. Smailer, Intern. Met. Rev., No. 5–6, pp. 231–264 (1979)
2   G. Hoyle, "Electroslag practices," Appl. Sci. Publ., London (1983)
3   H.-J. Fleischer, Stahl und Eisen 104, No. 15, pp. 727–735 (1984)
4   E. Plockinger, J. Iron Steel Inst. 211, pp. 533–541 (1973)
5   J. Kreyenberg, R. Thielmann, and P. Varhegyi, "Verbesserung des Gesamtausbringens beim Elektro-Schlacke-Umschmelzen mit hoher productivitat," Abschlubbericht des BMFT—Forschungsvorhabens, No. 3, pp. 70–71 (Apr. 1983)
6   D.M. Longbottom, A.A. Greenfield, G. Hoyle, and M.J. Rhydderch, Proc. 4th Int. Symp. ESR—Process, Tokyo, pp. 115–125 (1973)
7   P. Machner, Berg- und Huttenmannische Monatsch, No. 11, pp. 365–372 (1973)
8   L. Willner and P. Varhegyi, Arch. Eisenhuttenwessen, No. 47, pp. 205–209 (1976)
9   G. Bruckmann, "Beitrag zur Metallurgie des ESU—Verfahrens," Diss. TU Clausthal (1981)
10  H. Valentin, R. Diederichs, and H.J. Muller-Aue, "Einflub der Polaritat auf den ESU—Prozess," Interner Bericht Dew AG Krefeld (1969)
11  H. Jager, Berg und Huttenmannische Monatschefte 119, No. 11, pp. 439–447 (1974)
12  M. Kato, K. Hasegawa, S. Nomura, and M. Inouye, Trans. ISIJ 23, pp. 618–627 (1983)
13  T. El Gammal, Symposium on Welding and Heat Treatment Technology, Cairo (1974)

14  N. Nowack, K. Schwerdtfeger, and D. Krause, Arch. Eisenhuttenwes *53*, No. 12, pp. 463–468 (1982)

15  S.F. Medina and M.P. de Andres, Rev. Metal CENIM, No. 19, pp. 215–222 and pp. 271–282 (1983)

16  A. Mitchel and B. Beynon, Metall. Trans., No. 2, pp. 3333–3345 (1971)

17  R. Alberny, "Informationstagung «Gieben und Erstarren von Stahl», Verl., Bd. 1, pp. 320–388 (1977)

18  F.W. Thomas, "Elektrische Grossen im Schlackebad von Elektroschlacke—Umschmelzofen, Dissertation TU Hannover (1975)

19  T. El Gammal and M. Hajduk, Arch. Eisenhuttenwesen *49*, No. 5, pp. 235–239 (1978)

20  K.C. Mills and B.J. Keene, Int. Metals Reviews, No. 1, pp. 21–69 (1981)

21  G. Pateisky, Proc. of the 5th Int. Conf. on Vac. Metall. and Electroslag Remelting Processes, pp. 145–146, Munchen (1976), Hanau (1977)

22  J. Kreyenberg and K. Schwedtfeger, Arch. Eisenhuttenwes *50*, No. 1, pp. 1–6 (1979)

23  Schlackenatlas, "Dusseldorf," Verlag Stahleisen (1981)

24  Elektroschlacke, "Schlacke—Umschmelzen," Fachausschussbericht, No. 17 des VDEH, Dusseldorf (1984)

25  W. Holzgruber, Radex Rdsch., No. 3, pp. 409–421 (1975)

26  K. Bungardt, R. Diederichs, H. Freisenndanz, and E. Schurmann, DEW Techn., No. 12, pp. 157–177 (1972)

27  G. Lepie and H. Rellermeyer, Arch. Eisenhuttenwesen *37*, No. 12, pp. 925–934 (1966)

28  R. Gagnaux, "Erstarrung metallischer Schmeizen," DGM, pp. 261–291 (1981)

29  J. Delorme, M. Laubin, and H. Maas, "Informationstagung «Giessen und Erstarren», Luxemburg, T. 1, pp. 258–318 (1977)

30  R. Eggers and R. Jeschar, Arch. Eisenhuttenwes. *48*, No. 2, pp. 71–76 (1977)

31  K. Takahama and T. El Gammal, Third Germany–Japan Seminar, Verein Deutscher Eisenhuttenleute, Dusseldorf, Preprint No. 3, pp. 117–133 (1978)

32  A. Mitchell, Ironmaking and Steelmaking, No. 3, pp. 172–179 (1974)

33  W. Holzgruber and E. Plockinger, Stahl und Eisen *88*, No. 12, pp. 638–648 (1968)

34  G. Pateisky, H. Biele, and H.-J. Fleischer, J. Vac. Sci. Technol., No. 9, pp. 1318–1321 (1972)

35  H. Birck and F.W. Thomas, Digitale Regelung von Elektroschlacke—Umschmelz—Ofen, Hanau, Druckschrift der Leybold, Heraeus GmbH

36  Eintauchtiefenregler fur ESU—Anlagen, "Druckschrift der INTECO GmbH," Bruck a. d. Mur

# 43 Electroslag Remelting in Czechoslovakia and Its Future Development

## J. Wild and I. Kashik

The present level of development of electroslag remelting (ESR) of steel and alloys in Czechoslovak ferrous metallurgy is the result of the collaborative work among engineers, scientists, technicians, and economists that began in 1959. A history of ESR development in Czechoslovakia has already appeared in the Soviet technical literature [1, 2], including a description of the various development stages over a period of almost 30 years.

In the first stage, the goal was to design and construct an electroslag (ES) installation. In parallel with this work, research was started on developing basic ESR technology. Initially, automated welding equipment was used for this purpose, but this was later replaced by a general-purpose laboratory ESR installation. This stage of ESR development ended in 1968 when the first ESR shop was put into operation at the POLDI-SONP plant. The shop was equipped with ESR furnaces having sliding contacts, which were built at the plant.

Later, when it became necessary to increase shop capacity, new ESR installations were imported from the Soviet Union, giving the capability of producing ingots weighing 1.4, 2.5, 4, and 8 tons. This assortment of products was determined by the plant's capabilities to produce consumable electrodes with minimized recycling and without incurring additional expenses due to the need to rework ingots to the required dimensions.

The range of remelted steel types is given in Table 43.1. It is clear from the table that bearing, constructional, and chromium stainless steels make up the bulk of the remelted products. This type of assortment is determined by the economic requirements of Czech industry and is the most profitable for the plant as well as for the end users. However, the ESR shop capacity is not adequate to cover the needs for tool steel production.

This chapter is intended to describe our experiences in the development of ESR technology, which can be added to the voluminous material in this field published in the Soviet Union.

## The Working Principle of an ES Installation with Sliding Contacts

The aim was to build an installation with minimized power losses in the high-current electrode circuit and reduced power consumption, in comparison with other designs. The sliding contacts supply current from the transformer busbars

Authors' affiliations: J. Wild, POLDI-SONP, Kladno, Czechoslovakia; I. Kashik, IIChM, Karlschtein, Czechoslovakia.

**Table 43.1.** Types of remelted steels at the POLDI-SONP Plant

| Steel type | Percent of total production |
|---|---|
| Bearing steel | 40.6 |
| Constructional steel | 26.9 |
| Tool steel | 1.6 |
| High-speed steel | 2.1 |
| Stainless chromium steel | 27.3 |
| Stainless austenitic steel | 1.5 |
| Total | 100.0 |

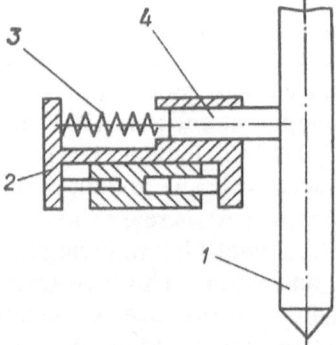

**Fig. 43.1.** Diagram of a single, current-carrying bar: (1) electrode; (2) base; (3) spring; and (4) contact pin.

**Fig. 43.2.** Overall view of the sliding contact assembly with a clamped electrode.

to the electrode by numerous contact bars that are capable of carrying currents of up to 10 kA. A diagram of the construction of an individual bar is given in Fig. 43.1. The sliding contact assembly consists of four hydraulically driven segments. Each segment is slotted to accept a number of spring-loaded bars that are held in contact with the cleaned surface of rolled, forged, or cast electrodes. An overall view of sliding contact assembly is given in Fig. 43.2. In order to use this type of current contact, it was necessary to build a mechanism for controlled electrode feed through the contact, into the mold. This device exerts pressure through a hydraulic mechanism onto grooved rollers that feed the electrode through the sliding contact and into the freezer mold. Such ESR installations with sliding contacts have been in use for 20 years in furnaces with an ingot capacity up to 1.4 tons. The power consumption for remelting bearing steels using ANF-6 slag is 1,240 to 1,330 kW-h/ton of steel.

## Technology of ES of Chromium–Nickel Stabilized Austenitic Stainless Steel

Oxidation of steel during remelting is undesirable. The oxidation potential of slags used in remelting is close to the potential of slags used in the reduction cycle of electric arc furnaces. However, it is high enough to oxidize the more readily oxidized alloying elements. The reaction of these elements with oxygen results in the formation of oxides that remain in the slag. This happens because the steel–slag system tends toward equilibrium.

These reactions may lead to nonuniformity in the chemical composition of the remelted steel, especially along the length of the ingot. In cases in which the steel–slag system is far from equilibrium, what usually happens at the initial stages of remelting is that the absolute value of the change of the reactive element concentration in the steel is rather high. This is also true for the change in oxide concentration in the slag. In the course of remelting, the system gradually approaches equilibrium, even though newly remelted alloy from the electrode, with the same concentration of reactive elements, continuously enters the system.

Reducing the oxygen partial pressure in the atmosphere above the slag layer was not found to be very effective in reducing or controlling the loss of easily oxidized elements. Therefore, element loss was prevented by maintaining the steel–slag system close to equilibrium by simultaneous additions of oxide ($TiO_2$, in our case) and reducing agents (Al) to the slag.

A mathematical model of the process was developed, which took into account the heterogeneous reactions and temperature variations in the steel–slag system. This model allows the chemical composition of a slag containing $TiO_2$ and aluminum to be calculated. The calculation is based on the condition that the steel–slag system is in equilibrium with the easily oxidized elements. The necessary reaction constants are determined from data obtained during test melting runs.

A process for remelting titanium-stabilized stainless steels has been put into commercial production. The process is used for remelting steels with a titanium concentration of 0.5% to 2.2% and a aluminum concentration no greater than 0.5%. Square ingots weighing up to 1.4 tons and measuring no more than $350 \times 350$ mm are produced. Round electrodes with diameters of up to 230 mm are used.

When this steel was remelted with ANF-6 slag and no additional measures were taken, the titanium loss was from 60% to 10%, depending on the titanium concentration in the initial alloy. The aluminum loss was considerably higher, although the aluminum concentration in the majority of these steels is low.

An ESR process using a $CaF_2$–$CaO$–$Al_2O_3$–$TiO_2$ slag deoxidized with aluminum was developed. One of the ingots was analyzed as-cast, and others were analyzed after hot working with six reduction cycles. The chemical analysis showed that the concentration of easily oxidized elements across the cross section does not show a significant difference between the ingot and the hot-worked, semifinished products. The main complications were caused by the nonuniform distribution of aluminum and titanium along the ingot length. However, by adopting the above measures to eliminate oxidation, it was possible to ensure, with a probability of 95%, that no systematic increase or reduction of titanium concentration along the ingot length occurred. The aluminum concentration increased in more than half of the ingots. However, these increases were minor, and the variation of aluminum concentration was no greater than ±0.04%; that is, in spite of the observed increase, the aluminum concentration still corresponded to the composition of this steel grade. It is interesting to note that the increase in the aluminum concentration was observed in the bottom end of the ingot, in the region where the process and the chemical reactions had been unstable.

The initial concentrations of aluminum and titanium and the average values and minimum and maximum values in five ESR ingots weighing 0.8 ton are given in Table 43.2.

The introduced process provides remelted steel with a predetermined composition with the required properties for subsequent processing and for end users' needs.

We would like to mention the results of another study in which a production technology provided high efficiency. ESR was introduced at the POLDI-SONP

**Table 43.2.** Aluminum and titanium concentration in five remelted ingots of austenitic stainless steel (12 analyses per ingot)

| Ingot number | Initial Ti (%) | Ti (%) after ESR | | Initial Al (%) | Al (%) after ESR | |
|---|---|---|---|---|---|---|
| | | Average | Range (max–min) | | Average | Range (max–min) |
| 1 | | 2.18 | 2.04–2.23 | | 0.37 | 0.31–0.39 |
| 2 | | 2.16 | 2.11–2.25 | | 0.35 | 0.32–0.36 |
| 3 | 2.8 | 2.16 | 2.10–2.25 | 0.49 | 0.34 | 0.29–0.36 |
| 4 | | 2.19 | 2.13–2.32 | | 0.35 | 0.32–0.40 |
| 5 | | 2.18 | 2.08–2.26 | | 0.39 | 0.31–0.39 |

plant at a time when all the steel was being cast as ingots. The electrodes for ESR were made from ingots of the appropriate weight and cross section; they were usually rolled (or forged, in the case of low-deformability steels). Somewhat later, however, several continuous casting installations were put into operation at the plant. Thus, electrodes that had previously been produced by hot working were replaced by continuously cast products. Since the cross-sectional dimensions of the cast electrodes were not compatible with the existing ESR process, a new method for clamping the electrode stub and for placing the electrode in the freezer mold had to be developed. In addition, it was necessary to introduce tighter control of the melting conditions. This work resulted in a one-third reduction of the cost of remelting and yielded the same steel quality as the previous method.

## The Effect of ESR on the Properties of a Hard Magnetic Alloy Used for Permanent Magnets

The average composition of the alloy is as follows: 0.04% C, 8.5% Al, 3.0% Cu, 15% Ni, and 25% Co (the balance is Fe). Permanent magnets for current applications are obtained by a casting technique followed by thermomagnetic treatment. The magnetic properties of the alloy depend to a large extent on the ingot structure, that is, on the degree of ordering and the size of the crystals. The structure of cast magnets (Fig. 43.3a) is satisfactory for most applications. For certain special applications, however, this structure needs to be improved. The problem of improving the ordering of the structure and increasing the crystal size was solved by ESR. Castings with a length-to-diameter ratio of less than three ($H < 3D$) were produced. These dimensions ensured that the heat flow was primarily through the bottom plate and also reduced the amount of heat absorbed by the freezer mold walls. This necessary improvement was achieved by using a special, highly conductive slag and selecting appropriate electrical parameters for remelting (Fig. 43.3b). The energy product[1] of magnets made from the remelted alloy was increased by 45% over that of cast magnets. Permanent magnets that were made from ESR alloy had exceptionally good properties.

The commercial use of ESR for producing high-quality carbon and alloy steels, which was started in Czechoslovakia in 1968, is gradually increasing. A further increase in production volume by 50% over present capacity is being planned in order to meet the needs for remelted steel in Czechoslovakia. For this purpose, modifications of existing ESR installations are planned, along with construction of new installations.

An analysis of the present state of ESR production of ingots weighing up to 8 tons shows that the process can be improved by employing higher productivity slags and perfecting process control, that is, by reducing the effect of control-

---

[1] The product of the coercive force and the residual magnetic induction [editor's note].

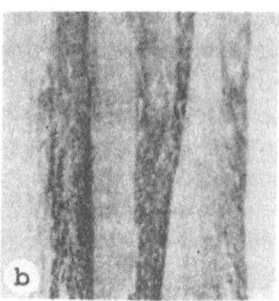

**Fig. 43.3.** Longitudinal templets of permanent magnets with 34-mm diameters produced by (a) conventional casting technology and (b) electroslag remelting.

lable factors on steel quality. Better control systems are expected to increase the productivity of some ESR installations at the POLDI-SONP plant.

The variety of types of ESR steels at the plant is more or less set at present, although there are no obstacles to an increase.

During the last few years, various new developments in electroslag technology (EST) have been tested, including processes for recycling scrap steel. This application of ESR is still in the initial development stage.

It is clear from this review that EST plays an important role in Czechoslovak ferrous metallurgy, notwithstanding the small production volume. Today, it is impossible to satisfy the needs for high-quality steel without the use of EST. We believe that technical progress in Czechoslovakia will be accompanied by a gradual increase in the production of remelted steel.

The history of the development and application of ESR in Czechoslovakia is marked by a collaboration between Czech and Soviet specialists, particularly from the E.O. Paton Electric Welding Institute. In this respect, the Czech specialists wish to express their gratitude to their Soviet colleagues.

# References

1  I. Kashik and I. Petrman, "The history of electroslag, remelting development in Czechoslovakia," Problems of Specialized Metallurgy 2, pp. 36–40 (1985)
2  B.I. Medovar, A.K. Tsykulenko, A.G. Bogachenko, and V.M. Litvinchuk, "Electroslag technology abroad," Naukova Dumka, Kiev, p. 320 (1982)

# 44 The Production of Large Electroslag Ingots with Low Oxygen and Aluminum Concentration from Steel Containing 12% Cr

## A. Suzuki and M. Okamura

At present, CrMoV steels are used in the manufacture of rotors for high- and medium-pressure turbines for fossil fuel power plants. During the last few years, steel containing 12% Cr has been used for turbine rotors used in modern, high-temperature steam generators.

Production of forgings from 12% Cr steel is complicated by problems with the chemical composition of the steel, which can be caused by reduction reaction products. Nonmetallic inclusions that are products of reduction accumulate in conventional ingots in the precipitation cone and act as defect initiators in the rotor forging; they can be detected ultrasonically [1].

In conventionally produced ingots of 12% Cr steel, axial shrinkage pipes develop, and a carbide eutectic and delta ferrite are observed in regions of chemical inhomogeneity [2]. Electroslag remelting (ESR) is considered to be the best approach to prevent these problems. This is because there is a high degree of refinement of the steel by the slag during remelting and because ESR produces ingots with a sound structure and minimal inhomogeneity.

However, an additional requirement for ESR steels used for rotors is their low concentration of aluminum and hydrogen. Therefore, hydrogen concentrations in both the electrode and during the ESR process were analyzed. Since the aluminum concentration in the ingot depends on slag composition, preliminary small-scale experiments were performed in order to select the appropriate slag.

The results of these investigations were used in the production of ESR ingots of 12% Cr steel with a diameter of 1.4 m.

## ESR Installation

A 50-ton ESR installation with a freezer mold diameter of 1.5 m that was built in 1970 was modified in 1983 for the production of 1.8-mm diameter ingots weighing 70 tons. The main characteristics of the modified installation are as follows: furnace type, single electrode with a stationary freezer mold; power supply, power transformer operating at 60 Hz with coaxial current leads; and process, liquid flux start, voltage control, automatic control of melting conditions, and measurement of melting rate by electrode weight sensors.

---

Authors' affiliation: Kobe Steel, Ltd., Japan.

## Special Features of the Installation

### Stationary Freezer Mold

There are three main furnace configurations: stationary freezer mold; movable freezer mold; and the furnace configuration permitting ingot withdrawal. The furnace is equipped with a stationary freezer mold, even though the largest ESR furnaces are configured with a movable freezer mold or ingot withdrawal. The stationary freezer mold provides a better ingot surface finish and a wider range for an acceptable variation of the melting parameters.

### Coaxial Current Leads

A coaxial construction is used to increase the power factor and to reduce magnetic interference.

### Control of the Remelting Process

To obtain a high-quality ingot, it is necessary to regulate the melting rate during the whole process. Since the remelting voltage depends on electrode penetration depth, the preset voltage should be maintained in order to regulate both the melting rate and the electrode feed rate. A weight sensor is used to measure the melting rate.

## Low Hydrogen Concentration

To obtain an ingot with a low hydrogen concentration, great attention must be paid to the production of the consumable electrode as well as to the ESR process. For melting 12% Cr steel, two arc furnaces with capacities of 60 and 100 tons were used. In accordance with the requirements, the steel was cast from both furnaces without the introduction of aluminum and silicon additives. In order to reduce hydrogen concentration, the steel from the 60-ton furnace was vacuum-degassed during casting, and the steel from the 100-ton furnace was vacuum-degassed in a ladle-refining installation. The hydrogen concentration in the steel was reduced from 30% to 60% by the vacuum treatment (Fig. 44.1). After correction of the chemical composition and the temperature, the steel was siphon-poured into the electrode mold under an argon atmosphere.

It was reported earlier that the hydrogen concentration in the molten metal pool during ESR depended on the partial pressure of water in the air. The data used in Fig. 44.2 were obtained for a low-alloy steel ingot melted in a freezer mold with a diameter of 800 mm [3]. Since a similar dependence was expected for the 12% Cr steel, ESR was performed under an argon atmosphere.

The slag is another source of hydrogen. The hydrogen concentration in molten slag is significantly lower than that in solid slag. Therefore, remelting was initiated using slag that had initially been melted in an electric furnace. This is the so-called liquid start method.[1] The hydrogen concentration in the slag and in the

---

[1]The authors use the term *hot start* [editor's note].

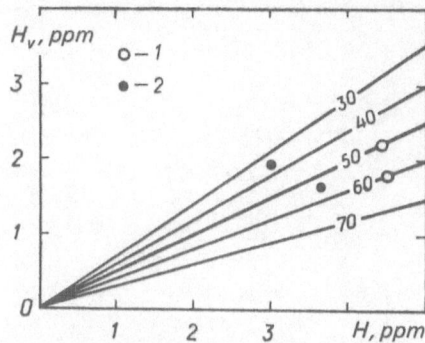

**Fig. 44.1.** Dependence of the hydrogen concentration ($H_v$) in vacuum degassed 12% Cr steel on the hydrogen concentration in steel before degassing ($H$): (1) degassing during pouring; and (2) degassing in a ladle refining furnace. The numbers on the curves denote percent dehydrogenization.

molten metal during ESR of 12% Cr steel in a 1.4-m diameter freezer mold is given in Fig. 44.3. Melting the slag significantly reduces its hydrogen concentration, and a liquid start provides a low hydrogen concentration in steel during ESR. The hydrogen concentration in the steel calculated from the model described in the literature [4] is given in Fig. 44.3 for a comparison with the experimental results, and also for a comparison of solid and liquid starts. Experimental data are in good agreement with the calculation for the remelting of two electrodes. In addition, as is clear from Fig. 44.3, the differences in hydrogen concentration in various electrodes were preserved in the remelted ingots. Thus, it is necessary to reduce the hydrogen concentration in the electrode alloy to obtain a low hydrogen concentration in the ingot.

It is evident from the calculated steel and slag hydrogen concentrations that, for a solid start, the hydrogen concentration must exceed that in the electrode alloy for the first five hours after initiation of melting. These results are confirmed by determinations of slag hydrogen concentration during remelting in a freezer mold with a diameter of 1 m.

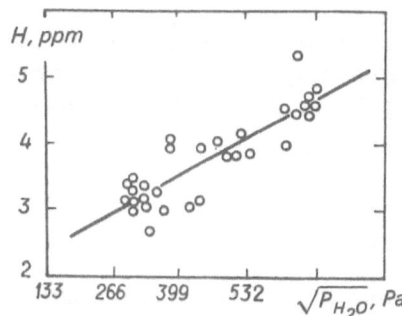

**Fig. 44.2.** Dependence of the hydrogen concentration in the molten metal pool on the partial pressure of water in the air (a 800-mm diameter freezer mold).

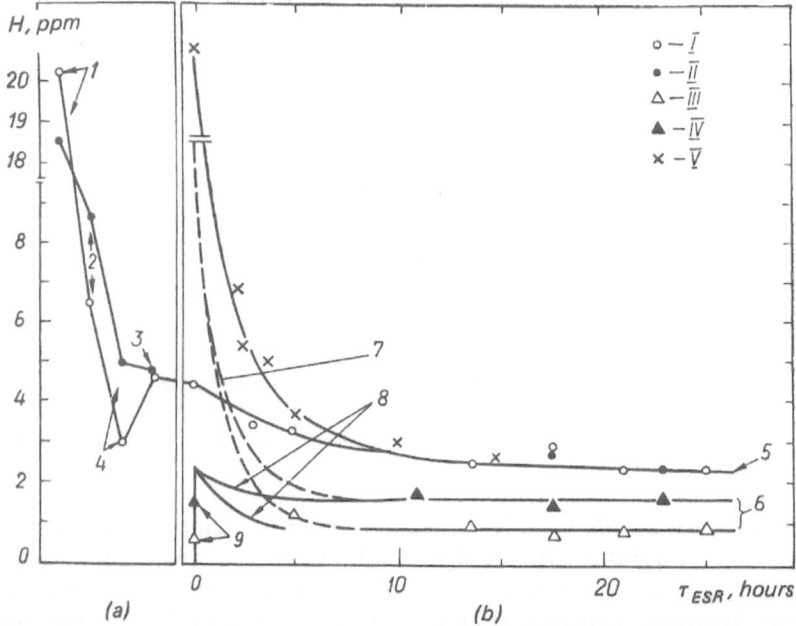

**Fig. 44.3.** Hydrogen concentration in the entire ESR cycle for 12% Cr steel. In the flux melting furnace (a): (1) before melting; (2) in the melt; (3) in the ladle; and (4) before pouring. In the electroslag furnace (b): (5) slag; (6) alloy; (7) and (8) calculated values for a liquid and solid start, respectively; and (9) in the electrode. In the slag and the alloy (I and III) (melting run A); in the slag and the alloy (II and IV) (melting run B). Runs A and B were done in a freezer mold with a diameter 1.4 m using a liquid start. (V) in slag (solid start, 1-m diameter freezer mold).

The hydrogen concentration of 12% Cr steel before and after ESR in a 1.4-m diameter freezer mold is given in Fig. 44.4. The hydrogen concentration does not increase even during the final stage of remelting.

## Low Aluminum Concentration

Since aluminum and silicon in the molten alloy interact with the slag, it is necessary to select the appropriate slag composition to provide a low aluminum concentration in the ESR ingot. The following reactions take place [5]:

$$3\,Si + 2\,Al_2O_3 = 4\,Al + 3\,SiO_2\,\ldots\,;\tag{1}$$

$$lg\,K = lg[a_{Al}^4\,a_{SiO_2}^3/a_{Si}^3\,a_{Al_2O_3}^2]$$

$$= -35840/T + 5.86,$$

where $K$ is the equilibrium constant; $a_{Si}$ and $a_{Al}$ are the activities of silicon and aluminum in the alloy; $a_{Si_2}$ and $a_{Al_2O_3}$ are the activities of $SiO_2$ and $Al_2O_3$ in the slag; and $T$ is the temperature in $K$.

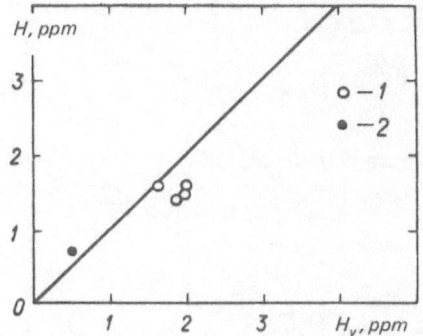

**Fig. 44.4.** Dependence of the hydrogen concentration in the molten metal pool on the hydrogen concentration in vacuum degassed alloy, before additional feeding: (1) experimental results; and (2) forged electrode.

The interaction parameters that were used for the calculations were as follows: $e_{Si}^{Si} = 0.11$, $e_{Si}^{c} = 0.18$, $e_{Si}^{Cr} = -0.003$, and $e_{Al}^{Al} = 0$.

A series of experiments was performed in order to choose the appropriate slag composition. As shown in Fig. 44.5, 400 g of 12% Cr steel were melted in a Tamman furnace in an argon atmosphere, and 30 g of slag was introduced to the molten steel. The steel/slag mixture was maintained at a temperature of 1,650 °C for 20 min after introduction of the slag, after which a sample of steel was obtained through a quartz tube. The 20-min waiting period is necessary in order to establish equilibrium between the steel and the slag. This length of time was based on the results of preliminary investigations.

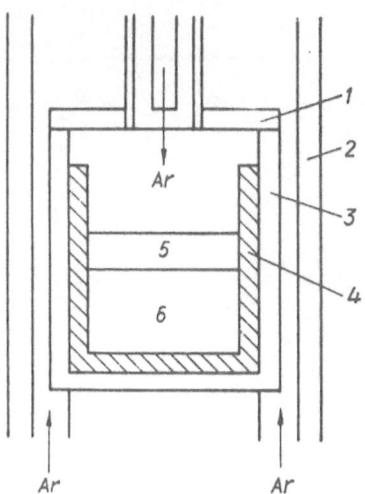

**Fig. 44.5.** Diagram of the experimental crucible: (1) quartz cap; (2) graphite heater; (3) graphite crucible; (4) sintered MgO crucible (46 × 38 × 100 mm); (5) slag (30 g); and (6) alloy (400 g).

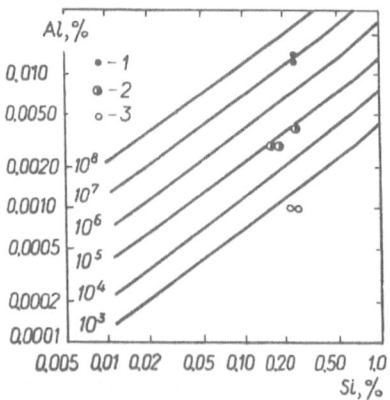

**Fig. 44.6.** Influence of the concentration of $SiO_2$ in the $CaF_2$–$Al_2O_3$–$SiO_2$ slag system on $a^2_{Al_2O_3}/a^3_{SiO_2}$: (1), (2), and (3) $SiO_2$ concentration of 0%, 3%, and 5%, respectively (12% Cr steel; temperature, 1,650 °C, crucible sample).

The effect of the silica ($SiO_2$) concentration in the slag on $a^2_{Al_2O_3}/a^3_{SiO_2}$ for the two slag systems $CaF_2$–$Al_2O_3$–$SiO_2$ and $CaF_2$–$CaO$–$Al_2O_3$–$SiO_2$ is given in Figs. 44.6 and 44.7, respectively. The value of $a^2_{Al_2O_3}/a^3_{SiO_2}$ decreases with an increased silica concentration in the slag.

The dependence of $a^2_{Al_2O_3}/a^3_{SiO_2}$ on the silica concentration in the slag is given in Fig. 44.8. For slags containing CaO, $a^2_{Al_2O_3}/a^3_{SiO_2}$ is higher than for slags that do not contain CaO. Therefore, CaO slags are less effective for producing steel with a low aluminum concentration. However, this does not call into question the efficient desulfurizing and deoxidizing capabilities of CaO-bearing slags. More-

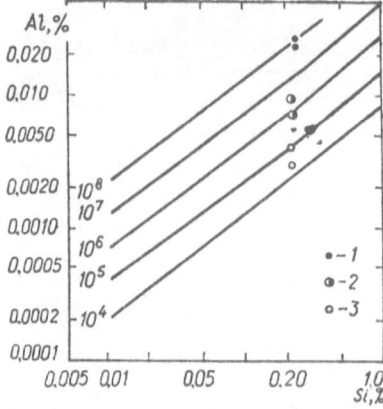

**Fig. 44.7.** Influence of the concentration of $SiO_2$ in the $CaF_2$–$CaO$–$Al_2O_3$–$SiO_2$ slag system on $a^2_{Al_2O_3}/a^3_{SiO_2}$: (1), (2), and (3) $SiO_2$ concentration of 0%, 5%, and 8%, respectively (12% Cr steel; temperature, 1,650 °C, crucible sample).

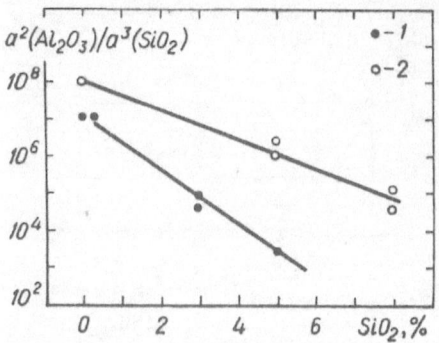

**Fig. 44.8.** Dependence of $a^2_{Al_2O_3}/a^3_{SiO_2}$ on the $SiO_2$ concentration in the (1) $CaF_2$–$Al_2O_3$–$SiO_2$ and (2) $CaF_2$–$CaO$–$Al_2O_3$–$SiO_2$ slag systems (12% Cr steel, crucible sample).

over, the use of a $CaF_2$–$CaO$–$Al_2O_3$–$SiO_2$ slag system containing 8% silica permits production of steel with an aluminum concentration of less than 0.005% (Fig. 44.7). This slag was studied in remelting experiments performed in a freezer mold with diameter of 480 mm. The results of these experiments are given in Fig. 44.9 and are compared with results obtained using a slag with no $SiO_2$. Through the use of a slag containing 8% silica, it is possible to produce steel containing about 0.001% aluminum and 0.05% silicon, whereas steel remelted with a silica-free slag has a higher aluminum concentration. Therefore, the slag containing 8% silica was chosen for ESR of 12% Cr steel in a freezer mold with a diameter of 1.4 m.

The electrode steel was produced in an 100-ton electric arc furnace and was subsequently treated in a ladle refining furnace. The concentrations of aluminum, silicon, oxygen, and sulfur in samples of molten steel that were taken dur-

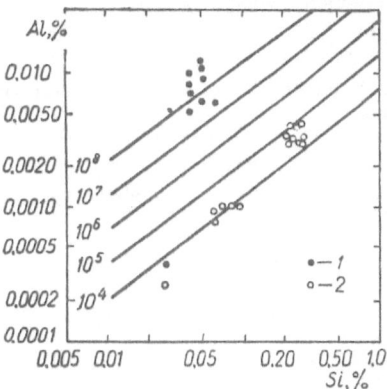

**Fig. 44.9.** Interdependence between Si and Al concentrations in ESR castings obtained using (1) a silica-free slag and (2) a slag containing 8% silica (12% Cr steel; temperature, 1,650 °C, a 480-mm diameter freezer mold).

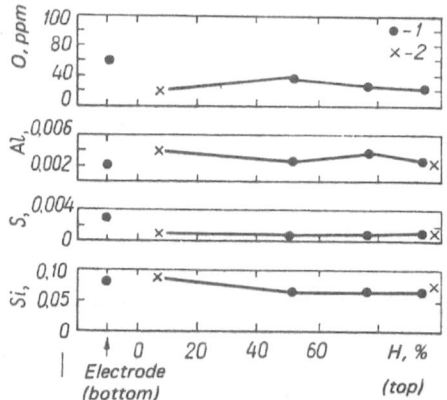

**Fig. 44.10.** Typical concentrations of aluminum, silicon, oxygen, and sulfur in the ingot during ESR in a 1.4-m diameter freezer mold: (1) molten metal samples; and (2) ingot surface samples.

**Table 44.1** Chemical composition of the top and bottom croppings from a rotor forging of 12% Cr steel obtained from a 1.4-mm diameter ingot (%)

|  | C | Si | Mn | P | S | Ni | Cr | Mo | V | Al | O |
|---|---|---|---|---|---|---|---|---|---|---|---|
| Electrode | 0.21 | 0.07 | 0.42 | 0.012 | 0.009 | 0.41 | 11.2 | 0.99 | 0.27 | 0.002 | 0.0087 |
| Top end |  |  |  |  |  |  |  |  |  |  |  |
| S | 0.22 | 0.05 | 0.43 | 0.014 | 0.003 | 0.42 | 11.32 | 1.00 | 0.28 | 0.001 | 0.0027 |
| MR | 0.21 | 0.05 | 0.43 | 0.013 | 0.003 | 0.41 | 11.29 | 0.99 | 0.28 | 0.001 | 0.0028 |
| C | 0.23 | 0.04 | 0.42 | 0.014 | 0.004 | 0.42 | 11.40 | 1.02 | 0.28 | 0.001 | 0.0027 |
| Bottom end |  |  |  |  |  |  |  |  |  |  |  |
| S | 0.21 | 0.06 | 0.41 | 0.013 | 0.002 | 0.41 | 11.22 | 0.97 | 0.27 | 0.002 | 0.0026 |
| MR | 0.21 | 0.06 | 0.41 | 0.013 | 0.002 | 0.41 | 11.33 | 0.98 | 0.27 | 0.001 | 0.0020 |
| C | 0.21 | 0.06 | 0.41 | 0.013 | 0.002 | 0.41 | 11.31 | 0.97 | 0.27 | 0.002 | 0.0022 |

Note: S, surface; MR, one-half radius; C, center.

**Table 44.2.** Nonmetallic inclusion content in top and bottom croppings from the rotor forging (according to ASTM E-45, method D)

| Electrode | Small sulfides | Large sulfides | Small alumina | Small spheroids | Large spheroids |
|---|---|---|---|---|---|
| Top end |  |  |  |  |  |
| S | 1.0 | 0.5 | 1.0 | 0.5 | 0.5 |
| MR | 0.5 | 0 | 0.5 | 1.0 | 1.0 |
| C | 0.5 | 1.0 | 0.5 | 1.0 | 1.0 |
| Bottom end |  |  |  |  |  |
| S | 0.5 | 0 | 0.5 | 0.5 | 0 |
| MR | 1.0 | 0 | 1.0 | 0.5 | 0 |
| C | 1.0 | 0 | 1.0 | 1.0 | 0.5 |

Note: S, MR, C, see Table 44.1; no large alumina or small or large silicate inclusions were observed

ing remelting, and also in samples taken from the surface of the solidified ingot, are given in Fig. 44.10. The results are satisfactory, and the differences between the two sets of samples are insignificant.

A second ingot was produced in the same freezer mold using the 8% silica slag. The electrode alloy was melted in a 60-ton electric arc furnace and was subsequently vacuum-degassed during casting. The ESR ingot was forged into a turbine rotor preform.

Croppings from the casting top and bottom ends were also analyzed. The chemical composition of croppings at three locations on the cross section are shown in Table 44.1. The nonmetallic inclusion content, determined according to the procedure in ASTM E-45 (method D), is given in Table 44.2. The re-melted steel is very homogeneous, with low concentrations of aluminum and oxygen.

## Conclusions

ESR was successfully used to produce ingots from 12% Cr steel for forged turbine rotors; the use of ESR solved earlier problems related to ingot quality, partic-ularly structural quality.

## References

1  A. Choudhury, R. Jauch, H. Löwenkamp, and F. Tince, Stahl und Eisen 97, No. 18, p. 857 (1977)
2  K. Akahori, S. Maeno, H. Kodama, N. Morisada, Y. Aikawa, Y. Ohmori, T. Ohshima, and K. Hisano, Tetsu To Hagane 69, P. 200 (1983)
3  T. Niimi, M. Miura, S. Matsumoto, and A. Suzuki, Proc. 4th Int. Symp. on Electroslag Remelt-ing Process, p. 322 (1973)
4  A. Masui, Y. Sasajima, N. Sakata, and M. Yamamura, Tetsu To Hagane 63, p. 318 (1977)
5  G. Pateisky, H. Biele, and H.J. Fleisher, J. Vac. Sci. Technol. 9, p. 318 (1972)

# 45 New Electroslag Furnaces Abroad

L.M. Stupak

The achievements of the modern steelmaking industry are largely due to the development of highly efficient methods of so-called secondary metallurgy, such as ladle (furnaceless) metallurgy, which have improved the quality of a large variety of steels. At present, the degree of molten steel purity is comparable to or even higher than that which earlier could be achieved only by remelting processes. Nevertheless, electroslag remelting (ESR) is still of great importance, as are other remelting processes. These technologies are used to produce high-quality steels having a high purity, density, and homogeneity. Increases in the purity of the initial alloy have yielded further improvements in the quality of ESR steel. This is currently the main development direction in world metallurgy.

The present state of ESR technology in developed capitalist countries is characterized by the continuous modification of existing electroslag (ES) furnaces and further improvement of ESR processes. Obsolete equipment is being replaced with newer designs. Additional new ESR shops are being organized in large companies. Some short descriptions of new ES equipment are given here.

## England

Two dual-position ES furnaces built by Leybold–Heraeus (West Germany) have been ordered and are being put into operation by Inco Alloys International Inc. (Hereford, England). These furnaces are intended for the production of ingots with a diameter of up to 630 mm and a weight of up to 6.5 tons from heat-resistant, nickel-based alloys for aerospace applications, including gas turbine components. The furnaces are equipped with second-generation computer systems that provide automatic control of the melting process. The control system automatically regulates the melting rate using feedback from the consumable electrode weight that is measured with a precision of 0.03% by a high-precision weight sensor. The furnace has a coaxial secondary circuit to reduce power losses. The furnace design allows for the application of the newest remelting technologies.

---

Author's affiliation: E.O. Paton Electric Welding Institute, Academy of Sciences, UkSSR.

## Italy

In 1983, at the Deltasider Plant (Aosta), the obsolete three-position ES furnace built by the Birlec Company was replaced by a universal, three-position furnace constructed by Leybold–Heraeus. The furnace is designed to allow electrode replacement during remelting and is capable of producing ingots weighing up to 44 tons in the central position using the ingot withdrawal method. Under production conditions at the plant, the maximum ingot weight is 35 tons and the maximum diameter is 1.2 m. At the other two positions, two ingots, weighing up to 11 tons, with a diameter of 850 mm, can be simultaneously produced in two stationary freezer molds. In these positions, it is possible to obtain ingots with square cross sections. The furnace is equipped with two thyristor-controlled transformers with a capacity of 25 KA; these can be connected in parallel when the central position is being used. In addition, the furnace is equipped with a coaxial bifilar secondary circuit, sliding contacts, a second-generation automatic control system using electrode weight sensors, a gas burner system for intensive heating of the replaceable electrode end, an air drying system, and a liquid slag system.

An ES furnace built by the Austrian Company FEV was put into operation at Nuova Italsider in Genoa. This furnace is of standard FEV design, providing electrode replacement, opposing motion of the freezer mold and electrode, etc. The ingot weight is about 10 tons.

A mono-bifilar installation for ES additional feeding of ingots weighing up to 100 tons, according to the TREST method, was put into operation at the Terni Company. The installation is used to process ingots after preliminary stream vacuum deoxidation. Casting of steel and additional feeding of an ingot are performed under conditions of minimized moisture content in the gas atmosphere above the slag layer. The installation is powered by either direct or alternating current and is equipped with an automatic control system.

## Japan

A commercial ESR furnace for the production of ingots for forging weighing up to 50 tons and an installation for ES additional feeding of slabs weighing up to 100 tons were put into operation at Kawasaki Steel Corporation in Mitsushima. The cost of both installations is $8.8 million. This equipment is intended for the production of extra-high-purity steel, which is initially subjected to a double refining step by the ASEA–SKF method. This extra-high-purity alloy finds application in the manufacture of submarine hulls, turbogenerator rotor shafts, rolls for high-capacity rolling mills, high-pressure pressure vessels, equipment for the nuclear power industry, etc.

The 50-ton ESR furnace at Kobe Steel, Ltd., was rebuilt in 1983. After modification, the furnace is capable of producing ingots of steel for rotor manufacture (12% Cr) weighing up to 70 tons, with a diameter of 1.8 m. The furnace characteristics are as follows: monofilar circuit, 60-Hz AC power supply; stationary freezer mold; coaxial cable; and liquid start. Special measures are taken to ensure a low concentration of hydrogen and aluminum in the remelted steel.

## United States

Two new ESR furnaces for the production of slabs weighing 20 tons from nickel- and cobalt-based alloys have been recently installed at the Huntington Alloys Corporation in Berno. A furnace for ESR of tool and other specialized steels into 19-ton ingots has been put into operation at Latrobe Steel Corporation in Latrobe, Pennsylvania. An ESR furnace for the production of corrosion-resistant steel ingots weighing up to 15 tons has been put into operation at the Joslyn Stainless Steel Corporation in Fort Wayne, Texas. All of these furnaces were built by the well-known Consarc Company. A fourth ESR furnace built by this company was installed at National Forge. This is a single-electrode furnace equipped with coaxial current leads and a system for ingot withdrawal from a stationary freezer mold. The furnace is used to produce 1.65-m diameter, 82-ton ingots for forging. The steel for remelting is manufactured in an electric furnace and is vacuum deoxidized. This is the second ESR furnace at National Forge; the first is a bifilar type furnace, which is capable of producing ingots with a diameter of 760 mm (30 in.) and a length of 1,010 mm (40 in.).

## Brazil

Due to the intensive development of its domestic steel industry using the latest techniques, in 1986 Brazil became the fifth largest steel producer among capitalist countries.

At the Electrometal Akos Finos S.A. Company (Sumare), an ESR furnace with a stationary freezer mold capable of producing 35-ton ingots was put into operation. In addition, a 10-ton furnace built by Consarc has been operating since 1973. The new furnace was designed and built in Brazil.

## South Korea

An ESR furnace with a capacity of 80 tons similar to the furnace that was built by Consarc and installed at National Forge was put into operation at the Hiyun International Corporation. There is also an ESR furnace at the KISKO firm that is used for the production of specialized steels.

## China

A three-phase ESR furnace[1] with six electrodes, for the production of 2.8-m diameter ingots weighing up to 300 tons was built and put into operation at the

Shanghai Heavy Machinery Plant. The furnace is equipped with three electrode holders, each of which is powered individually by a single-phase transformer, and an electrode replacement system. Two electrodes are held in each electrode holder. In addition, the furnace is equipped with an air drying system and operates with a liquid flux start. The ingot is withdrawn from the freezer mold as remelting proceeds.

## Yugoslavia

A second ESR furnace has been put into operation at the Zhelezarna Ravne Plant in Ravne-na-Koroshke. This furnace, which was built by the Austrian company Inteco, is equipped with a short stationary freezer mold and a movable bottom plate. The furnace is designed for electrode replacement, which permits the production of unique ingots weighing up to 36 tons with lengths up to 6 m and a diameter of 500 to 1,000 mm, or a cross section of $1000 \times 550$ mm. The transformer is rated at 3,250 KW-A, with a secondary current of 26 KA. The secondary current leads are bifilar; sliding current contacts are used.

## Democratic Republic of Korea (North Korea)

An ESR furnace for the production of 1.6-mm diameter, 90-ton ingots has been built and put into operation at the Dayanzhngige Manufacturing Association in 1982. The furnace has a single tower and a single electrode, with opposing motion of the electrode and freezer mold; the power transformer operates at 60 Hz and is rated at 7,500 KW-A at a current of 6,500 A. The furnace uses ANF-6 slag for remelting constructional steels used in the manufacture of large, highly stressed shafts, particularly steam turbine shafts.

The manufacturing association is planning to put a 180-ton ESR furnace into commercial production. Smaller ESR furnaces are also used at a number of manufacturing plants in North Korea.

## Conclusion

In the course of the modification of ESR equipment abroad, great attention is paid to improving remelting techniques and to automating the process by using computers. These measures are aimed at increasing the efficiency of ESR. The main problems are as follows: elimination of hydrogen in the ESR atmosphere, preliminary refining of electrodes, increasing the ingot length, and increasing the reproducibility of melting runs by using modern automatic control systems.

---

[1] See the chapter in this volume by D. Dzue et al., "A 200-ton electroslag furnace in the People's Republic of China" [editor's note].

# Bibliography

"The 30th Anniversary of Electroslag Remelting"
Paton, B.E., and Medovar, B.I.
*Electroslag Technology*, collection of papers devoted to the 30th anniversary of
   electroslag remelting, edited by Paton, B.E., et al., Naukova dumka, Kiev,
   1988, pp. 5–11
   Contains the history of the development, the present state, and future outlook
of electroslag remelting (ESR) technology in the Soviet Union and abroad.

"Contemporary Electroslag Crucible Melting and Casting, and Its Future Out-
   look"
Paton, B.E., Medovar, B.I., Marinski, G.S., Shevtsov, V.L., and Orlovski,
   U.V.
*Electroslag Technology*, collection of papers devoted to the 30th anniversary of
   electroslag remelting, edited by Paton, B.E., et al., Naukova dumka, Kiev,
   1988, pp. 13–19
   Contains a description of centrifugal electroslag casting and permanent mold
electroslag casting, which were developed at the E.O. Paton Electric Welding
Institute, and observations on the characteristics and applications of these new
processes, as well as a forecast of future development directions.
Number of illustrations: 5; number of references: 8

"EST as a Means of Improving the Design and Properties of Parts Used in
   Corrosive Environments"
Panasyuk, V.V., Katsov, K.B., Kovalenko, V.I., Rudenko, V.P., and Kuslitski,
   A.B.
*Electroslag Technology*, collection of papers devoted to the 30th anniversary of
   electroslag remelting, edited by Paton, B.E., et al., Naukova dumka, Kiev,
   1988, pp. 19–26
   Discusses the influence of nonmetallic inclusions on corrosion–fatigue fracture in corrosive environments and also problems connected with further improvements in electroslag steel quality. In addition, the authors point out the
advantages of ESR forgings and castings produced by centrifugal electroslag
casting for manufacturing high-reliability construction components for use in
corrosive environments and under high stress.
Number of illustrations: 3; number of references: 31

"Electroslag Remelting as an Efficient Way to Increase the Quality of Alloy for
   the Production of Wheels and Bearings"
Gasik, M.I., Proidak, U.S., and Gorobets, A.P.

*Electroslag Technology*, collection of papers devoted to the 30th anniversary of electroslag remelting, edited by Paton, B.E., et al., Naukova dumka, Kiev, 1988, pp. 27–31

Describes the high quality of steels used for the fabrication of wheels and bearings, which were melted in open-hearth and electric furnaces with subsequent refining using ESR.

Number of illustrations: 1; number of tables: 1; number of references: 4

"Electroslag Remelting in the Development of New Steels for Fabricating Parts and Bearings"

Levitin, V.S., Kropachev, V.S., and Zaharov, E.G.

*Electroslag Technology*, collection devoted to the 30th anniversary of electroslag remelting, edited by Paton, B.E., et al., Naukova dumka, Kiev, 1988, pp. 32–35

Describes the use of ESR technology for the development of various types bearing and tool steels used for manufacturing bearing components, and also iron- and nickel-based alloys used for manufacturing nonmagnetic bearings.

Number of tables: 1

"Electroslag Remelting of Steel Slabs for the Fabrication of Thin Strips"

Kamenski, Yu.M., Romanov, B.M., Drushinina, O.N. Perevalov, N.N., and Kakabidze, R.V.

*Electroslag Technology*, collection of papers devoted to the 30th anniversary of electroslag remelting, edited by Paton, B.E., et al., Naukova dumka, Kiev, 1988, pp. 35–37

Contains a description of electroslag production of slabs and their subsequent processing by rolling at the Serp i Molot plant (Moscow).

Number of references: 2

"Computer Automatization of ESR Furnaces"

Mahnenko, V.I., Gladki, E.D., Skosnyagin, Yu.A., and Zayats, V.I.

*Electroslag Technology*, collection of papers devoted to the 30th anniversary of electroslag remelting, edited by Paton, B.E., et al., Naukova dumka, Kiev, 1988, pp. 38–44

Contains a description of automatic control systems for ESR furnaces at the NKMZ Manufacturing Association (Kramatorsk), the Azovstal Plant (Zhdanov), and the Bolshevik Manufacturing Association (Kiev). These systems provide automatic control of the main remelting parameters (on the lower control system level), and also process control according to a mathematical model of thermal distribution in the solidified ingot (on the higher control system level).

Number of illustrations: 3; number of references: 3

"Electroslag Remelting for the Production of Tool Steels"

Yakovlev, N.F., Skrynchenko, Yu.M., Tishaev, S.I., Moshkevich, L.D., Pronorov, A.N., and Politaev, Yu.M.

*Electroslag Technology*, collection of papers devoted to the 30th anniversary of electroslag remelting, edited by Paton, B.E., et al., Naukova dumka, Kiev, 1988, pp. 45–49

Contains results of studies on the production of high-speed and tool ESR steels, which are used both as-cast and after forging. In addition, the influence of electroslag techniques on the quality of tool steels is demonstrated.
Number of illustrations: 3; number of tables: 1; number of references: 5

"Electroslag Remelting of Worn-Out Machine Tools"
Tkachuk, M.D., Tashlykov, V.I., Stetsenko, A.P., Seleverstov, O.G., Boiko, G.A., Zevakin, M.F., Saranchak, V.V., and Nebylitsin, L.E.
*Electroslag Technology*, collection of papers devoted to the 30th anniversary of electroslag remelting, edited by Paton, B.E., et al., Naukova dumka, Kiev, 1988, pp. 50–54
A technology for producing electroslag castings for forging dies using worn-out tools is described. It is also shown that the introduction of EST supplies 75% to 80% of the needs at the factory and yields savings 250 to 300 rubles/ton of dies.
Number of illustrations: 4

"Electroslag Casting of Preforms for Forging Dies"
Boiko, G.A., Miroshnichenko, V.A., Zykov, B.K., and Duplii, S.M.
*Electroslag Technology*, collection of papers devoted to the 30th anniversary of electroslag remelting, edited by Paton, B.E., et al., Naukova dumka, Kiev, 1988, pp. 54–57
Contains a description of the ESR shop at the Chernigov Automobile Part Plant and the Plant's work in the reuse of tool steel and flux.
Number of illustrations: 3; number of tables: 2

"Electroslag Technology at the 'Izhstal' Manufacturing Association"
Ponamarev, N.A., Zhdanovich, K.K., Trebov, N.P., Upshinski, E.A., Zaka-markin, M.K., Loiferman, M.A., and Lipovetski, M.M.
*Electroslag Technology*, collection of papers devoted to the 30th anniversary of electroslag remelting, edited by Paton, B.E., et al., Naukova dumka, Kiev, 1988, pp. 58–62
The results of studies of steel quality, mechanical properties, contamination by nonmetallic inclusions, concentration of dissolved gases, etc., are given. The influence of the ingot solidification rate on the length of internal defects is assessed, and optimal remelting conditions are proposed. In addition, a comparative analysis of two types of ESR tool steel is given.
Number of illustrations: 1; number of tables: 6

"Electroslag Remelting of Manganese"
Latash, Yu.V., Yakovenko, V.A., Lyuty, I.Yu., Butski, E.V., Bogdanov, S.V., and Kubikov, V.P.
*Electroslag Technology*, collection of papers devoted to the 30th anniversary of electroslag remelting, edited by Paton, B.E., et al., Naukova dumka, Kiev, 1988, pp. 62–66
Describes the studies performed by the E.O. Paton Institute and the Electro-spetsstal Plant on the ESR of electrolytic manganese. Results of studies of the

quality of ESR manganese and of the energy balance of the process are shown.
Number of illustrations: 4; references: 3

"Electroslag Remelting of Aluminum Bronzes Exhibiting the Memory Effect"
Larin, V.K., and Chernega, D.F.
*Electroslag Technology*, collection of papers devoted to the 30th anniversary of
electroslag remelting, edited by Paton, B.E., et al., Naukova dumka, Kiev,
1988, pp. 66–70
It is shown that ESR makes it possible to correct the electrode alloy composition and to adjust the thermal conditions for shape restoration of ESR alloy. The
use of cryolite- and fluorspar-based slags provides a high degree of refining. Due
to refining and favorable structural changes, ESR improves the strength and
thermomechanical characteristics of bronzes in comparison to refining and investment casting.
Number of illustrations: 2; number of tables: 2; number of references: 5

"Thirty Years of Electroslag Remelting Steel Production at the 'Dneprospets-stal' Plant"
Vodeniktov, I.G., and Gabuev, Yu.G.
*Electroslag Technology*, collection of papers devoted to the 30th anniversary of
electroslag remelting, edited by Paton, B.E., et al., Naukova dumka, Kiev,
1988, pp. 71–72
The contributions of the E.O. Paton Institute in the introduction of ESR for
mass production at the plant are noted. In addition, the characteristics of ESR
furnaces and the main projects at the plant are described.

"Electroslag Remelting at the 'Electrostal' Plant"
Klyuev, M.M., Kubikov, V.P., Pokrovski, A.A., Stepanov, V.P., and Fedot-kin, K.Ya.
*Electroslag Technology*, collection of papers devoted to the 30th anniversary of
electroslag remelting, edited by Paton, B.E., et al., Naukova dumka, Kiev,
1988, pp. 73–76
Reports on theoretical and practical developments in ESR at the Electrospets-stal Plant. In addition, future process developments are described, and predictions on improvements in steel quality are given.
Number of tables: 4

"New Developments in Electroslag Remelting at the Zlatoust Metallurgical
Plant"
Pokrovski, A.B., Hasim, G.A., Lazarev, V.I., Hrustalkov, L.A., Pozdnyakov,
V.A., and Kukartsev, B.M.
*Electroslag Technology*, collection of papers devoted to the 30th anniversary of
electroslag remelting, edited by Paton, B.E., et al., Naukova dumka, Kiev,
1988, pp. 76–78
Presents the results of research studies on ESR accomplished at the Zlatoust
Metallurgical Plant during a five-year period (1981 to 1985) and also a listing of
future work.
Number of illustrations: 2; number of references: 4

"The Fabrication of Small Preforms for Machine Components by Permanent Mold Electroslag Casting in Factories of the Ukrainian Light Industry"

Fishman, K.K., Zhalnin, A.V., Spivak, B.Ya., Yasko, N.I., and Orlovski, Yu.V.

*Electroslag Technology*, collection of papers devoted to the 30th anniversary of electroslag remelting, edited by Paton, B.E., et al., Naukova dumka, Kiev, 1988, pp. 79–80

Describes the production of small tools and machinery components weighing 0.5 to 5 kg by means of electroslag casting. In addition, the advantages of combined techniques are shown, such as ESC with pouring into ceramic molds, and electroslag investment casting, and the techniques of alloy proportioning and fabricating of disposable molds for producing of small castings are described. Data on the efficiency of using ESC in light industry are given.

Number of illustrations: 3

"Sectional Strip Mill Rolls with Electroslag-Cast Outer Linings Having a Variable Chemical Composition Across the Width of the Roll"

Matetcki, V.L., Nikolaev, V.A., Poluhin, V.P., Chernyh, V.V., Chepurnoi, A.D., Saenko, V.Ya., and Leshinski, L.K.

*Electroslag Technology*, collection of papers devoted to the 30th anniversary of electroslag remelting, edited by Paton, B.E., et al., Naukova dumka, Kiev, 1988, pp. 80–83

Discusses the effectiveness of using rolls with ESR linings. A comparison of the service properties of ESR and forged linings is given. Recommendations for the production of rolls that wear evenly are given. The recommendations are based on the analysis of roll wear profiles.

Number of illustrations: 2; number of references: 4

"The Introduction of Electroslag Technology at the MA 'Novokramatorsk Manufacturing Plant'"

Matsegora, E.A., Chepelev, A.T., Svistunov, G.N., Gavrishko, A.S., Kamalov, V.Z., Borovko; A.I., Grusko, Yu.A., Volkov, A.S., Fedorovski, B.B., Emelyanenko, Yu. G., Andrienko, C.Yu., and Maidannik, V.Ya.

*Electroslag Technology*, collection of papers devoted to the 30th anniversary of electroslag remelting, edited by Paton, B.E., et al., Naukova dumka, Kiev, 1988, pp. 84–86

The application of ESR steel increased the service life of critical components by 1.5 to 2 times, and also eliminated production rejects. Also describes development work on automatic control systems for the ESR-10G and 6ESR-20SV furnaces and planned future work on electroslag technology at the plant.

"Electroslag Casting at the Manufacturing Association 'Cheboksar Commercial Tractor Plant'"

Tsygurov, L.Z. Galkov, A.G., Ofitserov, E.M., Kuznetsov, V.N., Mironov, Yu.M., Kovalev, V.G., Atamanyuk, N.I., and Petelin, Yu.Yu.

*Electroslag Technology*, collection of papers devoted to the 30th anniversary of electroslag remelting, edited by Paton, B.E., et al., Naukova dumka, Kiev, 1988, pp. 87–90

Describes the ESR techniques that are used in the production of tractor components at the plant.

Number of illustrations: 3; number of references: 3

"Applications of the Electroslag Process for Producing Cutting Tools from Scrap Tool Steel at Factories of the Ministry of Machine Tools of the USSR"

Antonov, V.A., Miroshnichenko, A.G., Tkachuk, L.S., Odegov, E.V., Zherebetski, A.V., Linetski, V.B., Seroshtanenko, N.A., and Boiko, G.A.

*Electroslag Technology*, collection of papers devoted to the 30th anniversary of electroslag remelting, edited by Paton, B.E., et al., Naukova dumka, Kiev, 1988, pp. 91–94

Contains specifications for equipment in ESC shops for recycling tool steel scrap. Describes the new FISL-1 flux containing zircon concentrate and its influence on carbide inhomogeneity in R6M5 steel. As shown, the application of ESC H12M and H12MF steels increases the life of thread rolling tools by a factor of 2.5 to 3.

Number of illustrations: 3

"Electroslag Technology at the Manufacturing Association 'Kolomensk Heavy Machinery Plant'"

Yuzhanin, Zh.I., and Dubinski, R.S.

*Electroslag Technology*, collection of papers devoted to the 30th anniversary of electroslag remelting, edited by Paton, B.E., et al., Naukova dumka, Kiev, 1988, pp. 94–96

Describes the commercial introduction of electroslag casting at the Kolomensk Heavy Machinery Plant using a R-951UM laboratory installation and a commercial USh-108 furnace. In addition, the production of forging dies and rolls for continuous casting machines weighing up to 7 kg is described.

Number of illustrations: 2

"Electroslag Equipment at the 'Sibelektroterm' Manufacturing Association"

Zavyalov, V.G., and Pomeshikov, A.G.

*Electroslag Technology*, collection of papers devoted to the 30th anniversary of electroslag remelting, edited by Paton, B.E., et al., Naukova dumka, Kiev, 1988, pp. 97–99

Gives the specifications and history of improvement in furnaces used for producing ingots weighing from 10 to 20 tons and shows the benefits of commercial testing of furnaces in providing better service to end users of the equipment.

Number of tables: 2; number of references: 2

"The Use of Electroslag Technology in the Production of Electric Furnaces"

Volohonski, L.A., Kisselman, M.A., Nikulin, A.A., and Protokovets, E.G.

*Electroslag Technology*, collection of papers devoted to the 30th anniversary of electroslag remelting, edited by Paton, B.E., et al., Naukova dumka, Kiev, 1988, pp. 100–103

Gives the description of a remelting technique for bimetallic products and also gives examples of furnace components produced by ESR.
Number of illustrations: 3; number of tables: 5; number of references: 5

"Research at the Georgian Polytechnical Institute in the Field of Electroslag Technology"
Yakoboshvili, S.B., Mogilner, I.Yu., Haradze, D.M., Kobalava, G.Sh., Bikoev, G.G., and Hundadze, E.N.
*Electroslag Technology*, collection of papers devoted to the 30th anniversary of electroslag remelting, edited by Paton, B.E., et al., Naukova dumka, Kiev, 1988, pp. 104–106
Describes studies of electroslag technology conducted in the Welding Department of the GPI and at the Republic Center for Welding Techniques and Specialized Electrometallurgy.
Number of illustrations: 3; number of references: 3

"Developments in Electroslag Technology at the MRA 'NIIPTmash'"
Chernyh, V.V., Marchenko, I.K., Blohin, I.E., Molodan, G.A., Oleinichenko, V.I., Rudometkin, P.P., Lyubchenko, G.D., Bezhin, V.V., and Litvinov, A.M.
*Electroslag Technology*, collection of papers devoted to the 30th anniversary of electroslag remelting, edited by Paton, B.E., et al., Naukova dumka, Kiev, 1988, pp. 106–110
Contains the results of studies in the field of specialized metallurgy and comparative analyses of cast and forged steel and of ESR manufacture of highly stressed marine engine component, dies, rolling mill components, and other components. An economic analysis of the result of replacing forgings by ESR steel is given.
Number of tables: 3; number of references: 6

"Production of Large Ingots by Proportional Electroslag Casting"
Latash, Yu. V., Voronin, A.E., Biktagirov, F.K., Krutikov, R.G., Tynyankin, V.B., and Vasilyev, Ya.M.
*Electroslag Technology*, collection of papers devoted to the 30th anniversary of electroslag remelting, edited by Paton, B.E., et al., Naukova dumka, Kiev, 1988, pp. 111–114
Reports on the quality of large ingots produced by proportional electroslag casting. It is shown that due to favorable solidification conditions, and their fine dendritic structure, PESC castings have better properties than conventionally produced ingots and are similar in quality to ESR ingots.
Number of illustrations: 1; number of tables: 1; number of references: 4

"Large Hollow Ingots of Quasi-Monolithic Reinforced Steel"
Paton, B.E., Medovar, B.I., Shukstulski, B.I., Saenko, V.Ya., Chepurnoi, A.D., Lapin, V.V., and Chernyh, V.V.
*Electroslag Technology*, collection of papers devoted to the 30th anniversary of electroslag remelting, edited by Paton, B.E., et al., Naukova dumka, Kiev, 1988, pp. 114–116

Gives the data on the production of a hollow QMR ingot weighing 120 tons.
Number of illustrations: 2; number of references: 2

"A New High-Strength Quasi-Monolithic Reinforced Steel for High-Tonnage
BelAZ Dump Truck Beds Used in Mining"
Paton, B.E., Medovar, B.I., Saenko, V.Ya., Postizhenko, V.K., US, V.I.,
Medovar, L.B., Moiseenko, V.I., Chepurnoi, A.D., Shepotinnik, L.S.,
Gurov, N.A., Gizatulin, G.Z., and Mariev, P.L.
Electroslag Technology, collection of papers devoted to the 30th anniversary of
electroslag remelting, edited by Paton, B.E., et al., Naukova dumka, Kiev,
1988, pp. 116–121
Describes the properties of QMR sheet steel used for the construction of high-
tonnage BelAZ dump truck beds with a load capacity up to 180 tons and shows
the superiority of the new 22SMTYu-QMR sheet steel over 14H2GMR steel.
Number of tables: 1; number of references: 6

"Quasi-Monolithic Reinforced Spring Steel"
Saenko, V. Ya., US, V.I., Medovar, L.B., Vysotski, M.S., Gorbatsevich, M.I.,
Moiseenko, V.I., and Bondarkov, V.I.
Electroslag Technology, collection of papers devoted to the 30th anniversary of
electroslag remelting, edited by Paton, B.E., et al., Naukova dumka, Kiev,
1988, pp. 121–123
Gives the specifications and properties of 60S2-QMR spring steel. It is also
shown that application of this QMR steel for production of rear springs of the
MAZ-503A dump truck instead of monolithic steel increases spring service life
by 25% to 30%.
Number of tables: 1

"Economic Aspects of the Use of Electroslag Castings to Produce Machine Parts
at Kiev Factories"
Brechak, A.M., and Kryzhanovski, S.V.
Electroslag Technology, collection of papers devoted to the 30th anniversary of
electroslag remelting, edited by Paton, B.E., et al., Naukova dumka, Kiev,
1988, pp. 123–126
Discusses the economics of the production of components using resource-
conserving electroslag casting techniques. It is shown that electroslag casting is
more economical of material than forging. Lists the advantages of electroslag
castings and gives the results of calculations that show the high efficiency of ESC.
Number of tables: 2; number of references: 2

"Modern Electroslag Remelting Technology and Its Future Development"
Holzgruber, V.
Electroslag Technology, collection of papers devoted to the 30th anniversary of
electroslag remelting, edited by Paton, B.E., et al., Naukova dumka, Kiev,
1988, pp. 127–134
Compares ESR installations with stationary and short movable freezer molds,
and describes an improved control system for remelting. In addition, descriptions of ESR in a protective atmosphere and under high pressure, and electro-

slag casting of rolls for rolling mill, are given, as are data on the mechanical
properties of remelted steels.
Number of illustrations: 8

"Electroslag Remelting of High-Speed Steel in Hungary"
Sharvary, I.
*Electroslag Technology*, collection of papers devoted to the 30th anniversary of
   electroslag remelting, edited by Paton, B.E., et al., Naukova dumka, Kiev,
   1988, pp. 134–138
   It is shown that ESR significantly improves the service properties of steels,
especially of Lederburite steels.
Number of tables: 5; number of references: 3

"Electroslag Remelting of Specialized High-Alloy Steels"
Schlatter, R., and Bennani, A.
*Electroslag Technology*, collection of papers devoted to the 30th anniversary of
   electroslag remelting, edited by Paton, B.E., et al., Naukova dumka, Kiev,
   1988, pp. 138–145
   Describes an ESR installation with sequential electrode feed, which was built
by Leybold-Heraeus and is now used at Deltasider (Italy). Production techniques for high-alloy and constructional steels, tool steels, and heat-resistant
steels are discussed.
Number of illustrations: 4; number of references: 6

"Research in the Field of Electroslag Remelting at the University of British
   Columbia"
Mitchell, A.
*Electroslag Technology*, collection of papers devoted to the 30th anniversary of
   electroslag remelting, edited by Paton, B.E., et al., Naukova dumka, Kiev,
   1988, pp. 145–147
   Contains a brief review of studies in the field of ESR carried out at UBC since
1966.
Number of illustrations: 4; number of references: 6

"A 200-Ton Electroslag Furnace in the People's Republic of China"
Dzue, D., Haihon, L., Shuie, S., Dzyanhe, D., and Chuanlin L.
*Electroslag Technology*, collection of papers devoted to the 30th anniversary of
   electroslag remelting, edited by Paton, B.E., et al., Naukova dumka, Kiev,
   1988, pp. 147–150
   Gives the characteristics of the 200-ton electroslag furnace built at the Shanghai Heavy Machinery Plant. The quality of the ingots is assessed. It is also
emphasized that electroslag technology is expected to play an important role in
the production of large ingots for forging.
Number of illustrations: 3; number of tables: 6

"The Development of Electroslag Technology in Bulgaria"
Rashev, Ts.V.

*Electroslag Technology*, collection of papers devoted to the 30th anniversary of
   electroslag remelting, edited by Paton, B.E., et al., Naukova dumka, Kiev,
   1988, pp. 150–152
Describes the state of development of ESR in Bulgarian industry. Data are
given on remelting under pressure in a nitrogen atmosphere.

"The Development of Electroslag Remelting at the 'Huta Batory' Steel Mill
   from 1960 to 1985"
Krutsinski, M., Novak, A., and Strama, V.
*Electroslag Technology*, collection of papers devoted to the 30th anniversary of
   electroslag remelting, edited by Paton, B.E., et al., Naukova dumka, Kiev,
   1988, pp. 152–157
Describes the introduction of ESR technology at the Huta Batory steel mill,
where the first Polish ESR pilot plant was put into operation. The use of ESR
improved steel properties significantly. Presently, work is being carried out to
increase ESR steel production and to broaden the range of ESR products.
Number of tables: 3; number of references: 15

"The Experience of the 'Huta Beldon' Steel Mill with Electroslag Remelting"
Shvei, H. and Mista, S.
*Electroslag Technology*, collection of papers devoted to the 30th anniversary of
   electroslag remelting, edited by Paton, B.E., et al., Naukova dumka, Kiev,
   1988, pp. 157–163
The "Huta Beldon" steel mill specializes in the production of high-quality
ESR steels used to fabricate critical machine components and equipment. Specifications of ESR installations in current use are given. The proper combination
of installations with a movable and a stationary freezer molds allowed the production of ESR ingots with the required weight. The results of studies on the
influence of the slag composition on the quality and purity of remelted steel, and
on the power consumption, are given, as are data on the variations in the chemical composition of the slag and the degree of steel desulfurization during the
production of long ingots.
Number of tables: 5

"The Development of Electroslag Remelting at the 'Zhelezarna Ravne' Metallurgical Plant in Yugoslavia"
Rodich, J., and Schvaiger, M.
*Electroslag Technology*, collection of papers devoted to the 30th anniversary of
   electroslag remelting, edited by Paton, B.E., et al., Naukova dumka, Kiev,
   1988, pp. 163–168
Contains specifications of the unique ESR furnace built by Inteco (Austria)
that operates with sequential electrode feed and ingot withdrawal from a stationary short freezer mold. This installation was used to produce 1-m diameter, 6-m
long ingots that weighed 36 tons. The technique that was developed yielded
good chemical homogeneity along the ingot length.
Number of illustrations: 6; number of tables: 1

"Electroslag Remelting of Heat-Resistant Alloys: Thermal Balance of Melting and Alloy Chemical Homogeneity"
Domingue, J., and Yu, K.O.
*Electroslag Technology*, collection of papers devoted to the 30th anniversary of electroslag remelting, edited by Paton, B.E., et al., Naukova dumka, Kiev, 1988, pp. 168–176
Describes the application of ESR techniques at the Special Metal Corporation (USA). The advantages of ESR in comparison with vacuum arc remelting for the production of heat-resistant alloys used for aircraft engine components are discussed. In addition, the results of analyses of heat-resistant ESR alloys— Inconel 718 and Waspalloy—are given.
Number of illustrations: 2; number of references: 10

"Twenty-Five Years' Experience in Electroslag Remelting Development"
Rambo, J.B.
*Electroslag Technology*, collection of papers devoted to the 30th anniversary of electroslag remelting, edited by Paton, B.E., et al., Naukova dumka, Kiev, 1988, pp. 176–180
Gives a comparative analysis of ESR and VAR steels that are used in the aircraft industry for the fabrication of high-reliability components. Shows that ESR steel has better characteristics than VAR steel.
Number of illustrations: 4

"Parameters Influencing the Quality of Electroslag Ingots and Process Efficiency"
Thielmann, R., and Kreinenberg, J.
*Electroslag Technology*, collection of papers devoted to the 30th anniversary of electroslag remelting, edited by Paton, B.E., et al., Naukova dumka, Kiev, 1988, pp. 180–186
Contains studies of the interdependence between various process parameters, such as the melting rate, electrode penetration depth, slag composition and amount, cooling an additional filling conditions, specific power consumption, ingot surface finish, metal pool shape, shrinkage defects, and steel chemical composition.
Number of illustrations: 8; number of references: 36

"Electroslag Remelting in Czechoslovakia and Its Future Development"
Wild, J., and Kashik, I.
*Electroslag Technology*, collection of papers devoted to the 30th anniversary of electroslag remelting, edited by Paton, B.E., et al., Naukova dumka, Kiev, 1988, pp. 186–191
Describes the development of ESR at the POLDI-SONP plant in Czechoslovakia. Gives the specifications of ESR equipment at the plant, and describes remelting techniques for chromium nickel steel, tool and constructional steels, and also hard magnetic alloys. Future developments in ESR in Czechoslovakia are discussed.
Number of illustrations: 3; number of tables: 2; number of references: 2

"The Production of Large Electroslag Ingots with Low Oxygen and Aluminum
   Concentration from Steel Containing 12% Cr"
Suzuki, A., and Okamura, M.
*Electroslag Technology*, collection of papers devoted to the 30th anniversary of
   electroslag remelting, edited by Paton, B.E., et al., Naukova dumka, Kiev,
   1988, pp. 191–196
ESR was successfully used for remelting 12% Cr steel, which was used for
fabricating turbine rotor forgings. As a result of remelting, the chemical com-
position and the homogeneity of steel were significantly improved.
Number of illustrations: 10; number of tables: 2; number of references: 5

"New Electroslag Furnaces Abroad"
Stupak, L.M.
*Electroslag Technology*, collection of papers devoted to the 30th anniversary of
   electroslag remelting, edited by Paton, B.E., et al., Naukova dumka, Kiev,
   1988, pp. 197–199
Gives a brief listing of the specifications of new ESR furnaces that have been
built outside the USSR. The main development trends in ESR techniques and
equipment are also discussed.